DATE DUE

OG 24 '96			

DEMCO 38-296

SALVAGING THE LAND OF PLENTY

Also by Jennifer Seymour Whitaker

How Can Africa Survive?

SALVAGING THE LAND OF PLENTY

Garbage and the American Dream

Jennifer Seymour Whitaker

WILLIAM MORROW AND COMPANY, INC.
New York

 **This book is printed on recycled paper.**

Library of Congress Cataloging-in-Publication Data

Whitaker, Jennifer Seymour,
    Salvaging the land of plenty : garbage and the American dream / by
Jennifer Seymour Whitaker.
        p.   cm.
    Includes index.
    ISBN 0-688-10130-5
    1. Environmental policy—United States—Citizen participation.
2. Environmental protection—United States—Citizen participation.
I. Title.
HC110.E5W52   1994
363.72'8'0973—dc20   93-47490
                                                                      CIP

Printed in the United States of America

First Edition

1 2 3 4 5 6 7 8 9 10

BOOK DESIGN BY PAUL CHEVANNES

For my parents, Elizabeth McDonagh Seymour, and G. V. Menzies Seymour

Acknowledgments

I cannot adequately acknowledge the loving understanding and forbearance of my family and friends during the years I spent immured in the trenches with this book, or my gratitude that they were still there whenever I reappeared. I'm very fortunate and I owe you all.

So many people helped me along the way—starting with David Kellogg, whose infectious brainstorming sparked my original idea for the book, Flip Brophy, my agent and initial enabler, and Frank Mount, my first editor, whose enthusiasm for this project got me under way. At the Council of Foreign Relations, Peter Tarnoff's wholehearted support made it possible for me to carve out the time for research and writing while I continued to run the Committees on Foreign Relations program. The extraordinarily able backstopping of Pamela Gerla, Sharon Bially, and Brooke Russell enabled the Committees program to thrive during this time, as did the creative efforts of my friends and collaborators who lead the Committees in their cities. My Council colleague Ken Keller graciously served as a sounding board on scientific issues, shared his library, and gave an early draft of my chapters on science and technology an insightful reading. Shafiqul Islam heard me out on my basic problems with the discipline of economics, and also thoughtfully reviewed some of the manuscript at an early stage.

Other expert readers, whose counsel greatly enriched my thinking, included Steve Lewis, Roger Stone, Sheilah Mann, and Bette Fishbein. Greg Mark gave very generously of his summer vacation time, sagely checking the historical sections and contributing helpful edits to the whole manuscript. Throughout my research and writing process Fran Irwin generously shared her great expertise on the environment and public policy. During my research, the patient and ingenious library sleuthing and data collection of Pam Gerla, Evelyn Farkas, Fran Collins, Aimee Cotton, and Brooke Russell made possible my collection of materials in a variety of fields, and their thoughts and questions along the way added a lot to the mix. My invaluable summer intern, Sheldon Hirt, with a keen sense of the issues and considerable zeal, delved deeply on assorted fronts, and Danielle Kim, also on summer intern duty, worked with great energy putting a number of the disparate pieces together.

In the final stages, Deb Newitter scrambled even faster than I to meet the penultimate deadline, while Carrie Ciaccio put her considerable deductive powers to work tracking down loose facts that had eluded us heretofore and helping me cobble them all together. Ginny Parrott's computer talents and grace in a clutch also did a lot to get the manuscript on its way and looking good. Then my copyeditor and environmentalist extraordinaire, Kate Scott, submitted the whole text to invaluable substantive and stylistic scrutiny. For cheering me on and steering me shrewdly through the process of making my drafts into a book, special thanks are due to my editor, Bob Shuman. But *Salvaging the Land of Plenty* owes the most to my husband, Craig, who read it painstakingly, edited it with great skill, and made me feel that it was worth all the travail. Finally, I dedicate the book to my parents, who gave me life, love, curiosity, and the will to push onward.

Contents

Part I

~~~~~~~~~~~~~~~~

Progress
and Prodigality

Introduction: Culture and Conservation

No other nation on earth so swiftly wasted its birthright; no other in time made such an effort to save what was left.

—WALLACE STEGNER, *The Sound of Mountain Water*

WHILE I WAS working a few years ago on a book about Africa's development crisis, thoughts about the United States kept intruding. It was not that I saw many striking parallels between the two continents. Obviously they could not be farther apart in space, historical time, and material circumstances. But while writing about how Africans' economic choices were determined by culture, I found myself reflecting on the way our own attitudes about institutions were shaping, and limiting, our options.

As I thought about the dictates of culture—very broadly construed as shared attitudes about how society should work—one parallel did strike me. In the United States as in Africa, our ability to cope with economic and political dilemmas is eroding because our long-held convictions about social institutions no longer fit our current circumstances. In the richest of nations as in the poorest, efforts to grapple with the demographics, technologies, and power relationships of the late twentieth century

13

are often obstructed by the way culture leads people to approach problem-solving.

As I looked at Africa, one great obstacle to modern development there struck me forcibly. At the grass roots, Africans share a well-nigh universal conviction that communal stakes are primary, superseding those of individual owners, at one level, and of national government, at the other. In the mind of a businessman, the right of a brother or nephew to share in his take often overrides the need of his enterprise for profit or reinvestment. For members of a tribe or clan, their joint interest in getting hold of government revenues takes precedence over the claims of fellow citizens to equity and accountability at the national level. Thus group ties stifle entrepreneurship and at the same time distort economic management by government.

Americans, I realized, also resist the idea of their connection to a larger political entity. For us, however, it is not communalism but individualism that circumscribes our approach. The belief we have dearly held since the days of the Founding Fathers, that everyone can get ahead by their own effort if only government will stay out of the way, deprives us now of political tools we need to deal with economic and social change.

My comparison of the two vastly different societies ended right there. Yet it had prodded me to think about the *culture* of American economic development. Thus I began pondering the erosion of the United States' ability to solve societal problems before I focused on the erosion of our physical environment.

The range of deeply rooted problems we face in addition to environmental deterioration—in education, health care, housing, drug addiction, declining infrastructure—makes it clear that many of the essential remedies have to be applied societywide. Without a coordinated national effort, local and nongovernmental exertions in all these areas will continue to overlap, collide, and fall considerably short of what is needed, leaving us with a whole far less than the sum of the parts. Coming to grips with what ails us requires not only the guiding hand of government but also, more specifically, the leadership of the White House and Capitol Hill.

For a variety of reasons, the idea of such a common effort under government direction has appeared in recent years to inspire suspicion and distaste in many, perhaps most, Americans. After eight years under an antigovernment administration, many had thoroughly absorbed a presidential message they wanted to believe: that they shouldn't have to pay so much for government because they didn't really need it for anything except military and police protection. Moreover, the imperative for dismantling government, implemented through deregulation and administrative budget cuts and buttressed by the huge Reagan deficits, strengthened the antigovernment case by decreasing the effectiveness of federal oversight, implementation, and even auditing. It seemed to follow that government *shouldn't* do anything because it *couldn't* do anything.

As the Reagan era faded, I realized how far the Great Communicator's success resulted from his appropriation of ideas that have shaped the way Americans have viewed the world since our Revolution. For Americans facing a decline in our national share of the world's wealth (the first in our history), ancient verities resonated with renewed energy. The idea of virtuous individual entrepreneurship as opposed to invidious government intervention strengthened its grip on a people grasping for certitudes.

The power of this idea has undoubtedly increased the difficulties for Americans trying to organize common effort. When I contrast our attitude with that of the other industrial democracies, to my American eye the matter-of-fact accommodation to government authority in Europe and Japan seemed to make the implementation of policy enviably smooth. In particular, I felt, the European and Japanese acceptance of government *as the citizens' agent,* responsible for what they held in common, gave these countries a strong advantage in coping with change. As each nation struggled to adapt social and economic institutions to the hyperkinetic world of the late twentieth century, it helped to have someone minding the store.

At about this time, the wandering garbage barge, *Mobro,* hove into view. Like many other people, I was fascinated by the haywire quality of the *Mobro*'s voyage as it wended its way down the

Atlantic coast and through the Caribbean. Things seemed thoroughly out of control when the town of Islip on Long Island couldn't find a place *anywhere* to dispose of one bargeload of trash. Either we were about to burst our environment at the seams or American municipal government was in the final stages of decline and fall—or both.

In that moment the garbage crisis crystallized as an excellent focus for examining an American public policy watershed. Everyone has some garbage, and it matters to everyone else what they do with it. People, businesses, and government at all levels must grapple with it. They always have. Yet, as I realized increasingly while working on this book, the scale at which we must deal with the problem is entirely new.

Before long, the book outgrew my working title (now the subtitle), *Garbage and the American Dream*. What we usually mean by "garbage"—the refuse collected by municipalities—is only the tip of the iceberg. Household trash certainly adds considerably to our environmental woes. Yet its symbolic importance may be even greater: It represents the larger waste problem—all the detritus flowing from our production as well as our consumption. In this book, "garbage" usually means municipal solid waste—what homes and businesses set out to be collected. Yet the word is also used in a broader sense. Garbage writ large includes industrial waste; agricultural-, chemical-, and animal-waste runoffs; sewage; mining debris; and nuclear and military residues.

The cumulative effect of all this waste has changed our relation to the world we live in. I realized what this means only gradually, while researching and writing successive chapters about waste and pollution. The way it unfolded was a bit like the plot structure of a horror story, as each shocking saga or fearful fact was followed by another even worse one. I hope this will be the reader's reaction as well. We are being forced to face a new definition of pollution that is not just a matter of filth but of space: We're running short of room for both ourselves and our wastes. However, for most of us most of the time, the magnitude of our aggregate residues only becomes evident with the collapse of natural systems no longer able to absorb them.

As my research continued, garbage ceased to be a "case" I

was exploring for general truths about American public policy and took over the center of the book. Starting as an examination of American political culture within environmental policy-making, *Salvaging the Land of Plenty* grew into a study about the environment and public policy: how American political culture can best be mobilized to cope with the central problem human beings face during our remaining time on this planet.

My focus on waste, I also realized, was not entirely accidental. I have long been fascinated by ingenious ways to *save* things—time, space, materials, people. When I was a child, I was captivated by Ernestine Galbraith's account in *Cheaper by the Dozen,* of how her efficiency-expert mother ran her large family according to principles derived from time-motion studies. I found myself seeking the shortest trajectory between sink and cupboard as I dried the dishes, staking out the exact spot at which the school-bus door opened every morning, and peeling only the thinnest layer of skin to save more of the potato. At around the same time, the specter of invasive debris was firmly imprinted on my psyche by Thor Heyerdahl's story in *Kon Tiki,* the account of his 1947 raft trip from South America through the Pacific Islands. The idea that a primitive balsa raft hundreds of miles out in the South Pacific would run into floating plastic objects was deeply shocking to me in the innocent days of the 1950s.

As I worked through the argument of the book, my concerns about the American modus operandi on public policy and about what our economy is doing to our environment came together very naturally. Rooted in individualism and a mistrust of government, our adversarial style of politics has itself become enormously wasteful. Many of what we have counted as economic costs of our environmental controls result not from an escalating quest for environmental quality but from the rigid regulation and litigation endemic to a political arena where trust is at a minimum.

Moreover, as I pushed onward I realized how much the commonly accepted dichotomy between economy and environment distorts current public policy. Although we customarily view environmental goods and services within a much larger economic balance sheet, the human economy is in fact a subset of the environment. What we call "the environment" provides the necessary conditions for all economic life, from our energy and raw materials to human brainpower to the land, air, and water that make up our living space, and the sinks for our wastes.

∽

To understand the roots of our political mind-set, and the sweep of our economic/environmental evolution, I started at the beginning, with the ideas of plenty and freedom that have driven us always. Those who settled America found a land of incredible natural riches that made plausible the hope of wealth for every family and fostered the ideal of self-sufficiency. After their successful revolt against the King, the settlers' descendents sought to avoid government "meddling" in their economic lives, greatly favoring the neutral arbitration of the marketplace.

During the two centuries following our Revolution, Americans energetically exploited our resource endowment to produce more things for more people than any previous civilization could have imagined. Along the way we came to believe in our ability to solve most problems—including poverty—through technology and the market. Such hubris was understandable: By and large, we proceeded innocently, oblivious to any adverse physical consequences of our prodigious production. After World War II, however, enormous growth in population and science-based surges in industrial output (particularly in petrochemicals) magnified our impact on the environment exponentially.

Since the late 1960s we have been trying to clean up the mess we've made. For the most part, however, our progress has been halting and uneven. Some rivers and lakes are cleaner; ozone and smog have dissipated in some cities; sizable businesses manage toxic materials more cautiously. On the other hand, even in remote wilderness lakes more and more fish suffer from tumors and defective organs; coastal wetlands and shore areas are dying; perforations pock the ozone layer. We seem, in fact, to be moving one step forward and two back.

The advent of the garbage crisis after several decades of concern about pollution marked a new sort of public awareness that we have not been gaining on the problem. Despite various kinds of controls, the residues of our growth and development are vast, and they continue to increase. The concern about garbage represents a much larger public apprehension that we're running out of space for production and consumption as usual.

Now it has become clear that the way out lies in cutting back sharply on our use of materials. In seeking solutions, we have to look for measures that match the scale of the growing overload—not just that alleviate its symptoms. Even with municipal solid waste, although we need to make landfilling and incineration more effective, the answer does not lie in better garbage management—nor primarily in recycling. Our best means of coping with garbage and the overall waste crisis depend on reducing use and hence waste. Thus I focus at length on ways of cutting back the use of materials in both production and consumption.

However, this sort of comprehensive conservation will not happen automatically; in the present scheme of things progress is neither inevitable nor even likely. Neither technology nor the private sector can be relied on to deliver the sorts of change we need. Certainly, a wondrous array of technologies and processes is emerging to foster conservation, reuse, and reduction of waste. In addition, we are making strides in developing new sorts of scientific analysis that can help us better understand the consequences of our economic choices. Yet current economic incentives still favor exploitation rather than conservation of resources. Therefore, business, left to itself, is unlikely to bring the new technologies into full-scale commercial development anytime soon. So much American money and so many careers are invested in both resource extraction and profligate use that cutting way back will require extraordinary pressure.

Unfortunately, though, tightening up environmental regulations will not solve the problem either. The uniquely adversarial relationship between business and government in the United States greatly increases the costs and decreases the efficiency of government standard setting and oversight. Further, the task is simply too large to be accomplished by command and control mechanisms. Therefore, America really has no choice but to rely on the market to curtail environmental overload. It is the most efficient mechanism available to us for allocating and using resources—as well as the only one we trust.

Before the market can go to work on waste, however, we will have to use government policy to change incentives for the private sector. We are already experimenting with market mechanisms for pollution control through deposit systems and the sale of emissions permits. But the most effective way to curb waste is also the simplest. Substituting taxes on materials for many of the levies on income, sales, and property that currently fund government and common services will align market incentives with conservation.

Salvaging our environment will require Americans to override deeply embedded attitudes about production and consumption, ownership versus use, and property rights versus community needs. Ultimately, we will need to revise the way we determine value, to incorporate the present and future worth of both the resources we have until now virtually given away—including air and water—and the sinks for our waste. Yet for now we must use the means that are readily available. In redirecting market incentives through materials taxes we avoid unrealistic demands on our political institutions. Instead we rely on policies that draw on traditional American strengths.

In order to depict on one canvas where we have come from and where we must go, I have drawn on scholarship in a number of different areas. Encompassing the history of American consumption and waste, the ideologies that underpin our economic and political responses to pollution, current waste management practice in the United States, Europe, and Japan, and the economic dilemmas presented by environmental overload, this holistic approach has benefited from excellent studies in a variety of fields. Aided by these sources, I have endeavored to weave the social, economic, and political strands into a full panorama of our waste crisis and our society's coping mechanisms.

In proposing ways to change course I propose to engage critics of environmental regulation on their own ground. I have worked to convey the gravity of the problem without handwringing and moral reproach by showing the full gamut of waste that steadily permeates our land and water. And I have looked, first and foremost, for remedies that will foster economic efficiency as well as environmental quality. Ultimately, I hope my book will show that the two go in tandem—if we do the right things.

1

$\approx\approx\approx\approx\approx$

What's the Problem?

All things come from earth, and to earth they all return.
—MENANDER, *Monostikoi* (c. 342–290 B.C.)

What goes around comes around.

DURING THE PAST century we have banished limits on production, yet a new sort of scarcity is changing the way we live. We are running out of space on this planet for our waste. In the past hundred years we have learned how to produce an apparently endless stream of goods, and to sustain populations grown fourfold. Now, however, our production process has outpaced the absorptive capacity of our habitat. That fact is likely to change our lives even more than the material revolution that created the glut.

That we have a national garbage problem has become a staple of ordinary conversation for many Americans. Per person we throw away twice as much household trash as the Europeans and the Japanese. For most U.S. communities, disposal costs are rising, landfills are overflowing, underground water is tainted, and nobody wants incinerators in their neighborhoods. As the press bombards us with tales of towering landfills and white elephant incinerators, out of sight is no longer reliably out of mind.

Many big bills for the maintenance of our lifestyle are coming due at the same time. Industrial plants are surrounded by fields of poison residues; rain washes toxic fertilizers, insect spray, weed killers, and manure into rural rivers and streams; aging nuclear weapons plants sit side by side with radioactive ponds and collected debris. The waste involved in everyday life and commerce often has more serious consequences than industrial dumping or the occasional catastrophic accident. Twice as much oil leaks into American soil around underground fuel tanks every six months as that which spilled from the tanker *Exxon Valdez* into Alaska's Prince William Sound, and seventeen times as much is dumped into the Mediterranean every year from the flushing of tankers and minor accidents. Human and commercial wastes washing through the sewage system imperil coastal wetlands and urban water supplies. Stream diversion, development, and farm runoffs threaten the existence of freshwater fish species in lakes, rivers, and streams throughout America.

Recently, widespread public recognition of our waste problem has changed the nature of environmental concern from a problem we experience in periodic bursts of fashionable angst to a crisis that will not go away. From the latter years of the nineteenth century through much of the twentieth, most Americans and Europeans enjoyed a passionate and innocent infatuation with technological progress. For the first time in human history Western nations were creating wealth much faster than people. When progress at times gave rise to new problems, as in the first large-scale U.S. municipal garbage crisis in the early 1890s, technology could be summoned to solve it. Given the impressive growth of science-based industry, this faith in technology ruled material life even through systemic disasters like the Great Depression. Following World War II, an extraordinary burst of material growth buttressed the credo yet again.

Starting with Rachel Carson's powerful assault on pesticides in her 1962 book, *Silent Spring*, Americans' trust in the power of technology to smooth away material problems sustained several major blows. During the late 1960s and early 1970s a series of alarms about dwindling resources and threats to health from polluted air and contaminated water spurred a potent new environmental movement in the United States that ultimately led to pathbreaking national legislation aimed at pollution control.

By the mid-seventies, however, activism and attention brought on by the 1973 OPEC oil embargo had generally diminished, just as the aura of the 1970 Earth Day celebration had receded into the fading counterculture. By the early 1980s, with the price of energy plunging and OPEC members fighting to sell their oil, the idea of conservation seemed dated. A decade after the U.S. Clean Air and Water acts of 1970 and 1972, environmental activists still deplored the do-nothing stance of the Reagan Administration, but for the general public, pollution abatement posed no problems that money and expertise could not fix. Viewed from a national perspective, environmental anxieties appeared likely to ebb and flow, like other social fashions, even though grass-roots groups were gearing up for local battles.

Although it had been building over many years, the crisis seemed to arrive without warning. Quite suddenly, nature intervened, shifting the terms of the relationship between people and their habitat.

The seeming abruptness resulted from the fact that exponential growth accelerates sharply as it crosses a fixed limit. Toward the end of the 1980s, it became apparent that wayward components of aerosol sprays, chlorofluorocarbons, or CFCs, were punching holes in the earth's ozone layer and that emissions from burning fuels were probably changing our climate. In addition, for some years reports of toxic spills and seepage had been invading public consciousness. For the first time, Americans expressed fears about damage to the whole environment, destruction that might be greater than the effects of the myriad smog, acid rain, sewage, or chemical dumping problems they had always viewed as regrettable but inevitable by-products of industrial growth.

The "garbage crisis" became the most tangible expression of these fears. People were beginning to suspect that our production and consumption sprees might resemble the adventure of Walt Disney's sorcerer's apprentice, who knew enough of his master's magic to enlist battalions of animated mops and pails for his cleaning chores but not enough to make them stop or to staunch the floods they flung at him. We didn't know how to make all the extra *things* disappear. Concerns about garbage disposal, landfill and incinerator emissions, groundwater contamination, and toxic waste dumps grew in the broader public.

Some experts have characterized these anxieties as hysteria. They argue that popular apprehensions are misplaced and that compared with other risks (including household accidents and the perils of the road, smoking, and fatty diets), these environmental threats are just fashionable bogeys posing significantly less peril to life and limb. According to this view, the preoccupation with cancer has derailed rational discourse about actual dangers from other quarters. People are ignoring scientific findings about the real risks to health under probable conditions for human exposure.

Others argue that though garbage itself is a periodic public fixation, much of what worries the American public right now can be readily managed. As far as household trash is concerned, the crisis is a product of rising standards and local politics. If we did not worry so much about contamination of air, land, and water by the old easy ways of dumping and burning, there would be no garbage glut, only more landfills and incinerators. In a lively recent book, *Rubbish!*, garbage archaeologist William Rathje, who has undertaken excavations in a number of landfills, takes aim at what he considers current myths about the dangers of plastic, polystyrene, and fast-food packaging, while boosting what he sees as a relatively benign process of garbage mummification in landfills. Alternatively, proponents of industry's preferred option, incineration, assert that once the recycling craze has spent its momentum, a rational public will opt for new "waste-to-energy" technologies for "nonpolluting" incineration.

Many of these observations are valid, but they largely ignore the real source of popular anxiety. After a decade of concern, we manage our municipal waste much more cleanly. Nonetheless, it is not the Styrofoam cup itself that people recoil from, but the longer-term threat lurking behind it. For many ordinary Americans, household trash represents a much larger environmental quandary, way outstripping the issue of priorities. "The garbage crisis" has become shorthand for the system's overload. The overflow of household refuse became a metaphor for environmental saturation as trash-laden barges floated around the world trying to dump refuse and incinerator ash from Philadelphia and Long Island; landfills leaked toxic liquids into groundwater; citizens vetoed municipal funding for incinerators. At the same time, massive contamination over decades by civilian and defense industries as well as agriculture was unmasked, revealing costly and in some cases irreversible damage from toxic and radioactive residues.

With air, land, and water apparently no longer able to absorb the residues of the good life, environmental problems took on a new dimension. As long as industrial production led to "pollution," or "poison," coping with it implied as remedy some sort of cleanup or isolation. But when the focus shifted to limits in the absorptive capacity of the planet, "pollution" became translated into "waste." If the globe has only a certain amount of room for all the stuff that we cast aside, that fact becomes a compelling argument for absolute limits on materials flows. Then technologies for sequestration of toxics are not enough.

Two centuries ago, Thomas Malthus, in "An Essay on the Principle of Population," contended that poverty and famine were inevitable, since population increases geometrically and the means of subsistence arithmetically. Since then, people have ardently debated whether humans could overrun the earth's capacity to provide for them. The idea has always exercised considerable power over people's imaginations. But each time the Malthusian apocalypse failed to arrive, Malthus's insights were further discredited. A current neoconservative backlash against environmentalism sounds a variant of this refrain.

Writing in this vein in the June 1993 inaugural issue of the business-oriented environmental journal *ECO*, Gregg Easterbrook characterized public anxiety about the degradation of the biosphere as a sort of hysteria. He also suggested that the environmental movement wants to "hush up" good news about gains against pollution, partly to safeguard the momentum of its own fundraising. Ignoring the fact that environmental headway in the United States has resulted almost entirely from the pressure of grass-roots groups, he argued that public alarmism about human impacts on nature impedes real progress in coping with these impacts. What he seems to be suggesting is that only when the boy stops crying wolf will the community begin to keep watch. Even *Garbage* magazine, founded in 1989 as concern crested, three years later was expressing exasperation with over-heated environmentalists, as, for example, in the October/November 1992 issue, when Publisher Patricia Poore characterized the campaign against disposable diapers as "environmental vigilantism."

In fact, however, the holes in the ozone layer provide the first incontrovertible evidence that Malthus may have been right. Because he could not foresee technology's power to compensate for human pressures on the earth's physical limits, during the past hundred years Malthus's prediction has appeared dead wrong. In his own times, before the Industrial Revolution, the equation seemed simpler. In many parts of eighteenth-century Europe (as in Africa and other poor Third World areas today), farmers had denuded many of their forests for fuel and in some places had worked the soil to exhaustion. As towns and industries grew, however, and the link between sustenance and the land became progressively looser, humans seemed to be divesting themselves of nature's shackles. By the latter years of the last century, people in the West were beginning to believe that technology could provide all that was needed to just about everyone.

In reality, they hadn't seen anything yet—an extraordinary surge in growth awaited in the mid–twentieth century. While U.S. population, spurred by the postwar baby boom, shot up from 150 million to 250 million between 1950 and 1990, production spurted even more spectacularly, doubling and then redoubling during the 1950s and 1960s. By 1980, industrial capacity in North America and Europe was two hundred times as large as in 1900. Our success in bringing so many more goods to so many more people might have seemed to put to rest Malthusian fears. Instead of solving the problem of overpopulation, however, the production revolution intensified demographic dilemmas. Accelerated output of goods, compounded by exploding populations, brought vast new strains on land and water.

As a result of our extraordinary population increase, in the past forty years human beings have altered the portion of the earth and its atmosphere that supports life, the biosphere, more than in all preceding human history. In their seminal account of human-induced biospheric change, *The Earth as Transformed by Human Action*, B. L. Turner and a group of physical and social scientists carefully detail the accelerated changes in the environment after the Industrial Revolution. Before that, and except for some localized areas of siltation, salinization, alluviation, and extensive but partly reversible deforestation, they assert that human "impacts were few and essentially negligible." From the Industrial Revolution to the present, however, and particularly since World War II, human population growth and economic development have changed the environment enormously. The authors produce massive evidence of deforestation, loss of species, declining water supplies, and growth of population, as well as huge increases in chemical and metals emissions.

The study acknowledges at the outset that the most basic and far-reaching changes in the earth's environment over the planet's four-billion-year history have been driven by nature rather than man. The tectonic and geomorphologic forces that created the biosphere itself, moved continents and oceans around, fostered the evolution of millions of plant and animal species, and from time to time radically altered climates continue to act on the biosphere today. The last major climatic shifts, however, happened over four thousand years ago, at the time that our species began to make a significant mark on the planet.

At that point humankind joined the other natural and cosmic forces exercising transforming power over our environment, but for most of subsequent history our impact was still minimal. People modified the land and its populations of flora and fauna through fire and the felling of forests, irrigation and cultivation, as well as through the slaughter of predators. It was only three hundred years ago that the velocity of the human intervention began to really accelerate.

Throughout the last three centuries the pace has quickened sharply. In the eighteenth century the widespread transformation of European agriculture from subsistence to surplus that preceded industrialization deepened the human imprint on landscapes and species. Then, in the nineteenth century, with the first phase of the Industrial Revolution, fossil fuel–based and mechanized agriculture and manufacturing began to seriously alter the flows of energy and chemicals. Increasing per capita consumption, based on the extraction of energy and minerals and the creation of new chemicals, continued and accelerated in the twentieth century.

Most of what we have done to the biosphere we have apparently done very quickly. On the population side, numbers of people grew sluggishly through most millennia of human existence, reaching a billion only around 1812. Since then, however, the intervals between billions have rapidly decreased, from 115 years to 35 to 15, and now 11. While most of the net loss of forests and vertebrate species had already occurred by the early years of this century, 50 percent of the release of carbon into the environment over the last three hundred years has happened since 1950, and emissions of lead, sulfur, phosphorus, nitrogen, and carbon tetrachloride, which were negligible one hundred years ago, doubled after World War II. Our global use of fresh water, which rose from about 100 cubic kilometers at the start of the Industrial Revolution to 1,800 cubic kilometers in 1940, has doubled since then to today's level of approximately 3,600 cubic kilometers.

Turner's study points out a much less obvious impact from the human release of energy in the form of heat, the consequences of which we do not yet understand. This release, while equal overall to only 1/10,000 of the heat produced by solar radiation to the earth, equals the sun's heat during the winter months in major cities of North America and Europe. The study also notes that "high temperature industrial emissions alone now multiply the annual natural releases of arsenic by 3, of cadmium by 7, of mercury by 10, and of lead by 25." To these must be added great quantities of substances not present in nature; the output of synthetic organic chemicals has increased by almost 3,600 percent since World War II. The suddenness of these changes has been most dramatically manifested by the damage to the ozone shield, an event not predicted by any of the scientific monitoring and measuring mechanisms we have come to rely on.

The wastes from all this human activity, garbage writ large, are absorbed—as they have been historically—into the land, and from there into streams, rivers, lakes, and oceans. The open oceans, because of their vast diluting capacity, appear relatively less affected by the impacts of the past several centuries, but inland and coastal waters, often the primary dumping areas for industries and residential areas as well as the receptacle for farm and other diffuse sources runoff, have been under severe siege for the past several decades. Between 1950 and 1980, the volume of river water polluted by wastes trebled. Now, while water pollution abatement is beginning to revive rivers like New York's Hudson, it appears we may have been only pushing the problem farther out, as dead zones appear in coastal waters like those near the continental shelf of the western Atlantic. And while we have made some progress in capturing pollution from major industries and waste treatment plants, we can now begin to glimpse another vast universe of waste emanating from farms, construction, small business, and spreading development.

In the 1990s the notion of this planet and its enveloping atmospheric layers as a closed system has become incorporated into our shared imagery, but our understanding of what we have wrought still lags far behind our ability to change our world. We are only beginning to perceive the complex links between carrying capacities of natural systems and human activity. Yet our dawning awareness of these fundamental interconnections has made "what goes around comes around" an apt maxim for our time.

Whereas we previously focused on resources and pollution as separate concerns, the two are now connected in the idea of waste. We have begun to realize that all used-up resources eventually become "waste," and all waste that the system cannot absorb is potentially polluting. The rate at which materials are converted to waste is striking. Economist Robert U. Ayres estimates that only 6 percent of minerals and renewable materials extracted each year are embodied in durable goods; the other 94 percent are converted into waste residuals. Thus, the best way to curb both pollution and waste is, ultimately, to curtail our diffusion of resources.

Throughout history, all settled peoples have grappled with wastes. In the Stone Age, cave dwellers simply moved on when the debris piled too high. As historian Martin Melosi and others point out, it was not uncommon in the ancient world for people simply to drop their garbage on the floors of their houses or on the streets outside. Building over the debris, the inhabitants of

Troy and cities in North Africa gradually raised the levels of their floors—sometimes as much as twenty inches in the towns. In most places, the degree of sanitation reflected the prosperity of a neighborhood, as public garbage collection was rare and the wealthy deployed servants to clean their environs.

Some early civilizations also cared enough about cleanliness to devise garbage and sewage systems having much in common with those in modern cities, as well as various sorts of public and private refuse collection. Four thousand years ago, cities in China, India, Egypt, and Israel passed ordinances regulating the disposal of municipal waste. In the Indian city of Mahenjo-Daro, homes were constructed with built-in rubbish chutes and trash bins, attended by a scavenger service. The ancient Mayans and others carted their garbage to the dump. Residents of the Punjab city of Harappa furnished their homes with bathrooms and drains, while ancient Babylonians also enjoyed sewage systems with drains and cesspools. In Crete the royal palaces boasted bathrooms connected to trunk sewers, and eventually authorities set aside dumps for organic waste.

In China in the second century B.C., municipal services in major cities included both "sanitary police," who carted off human and animal carcasses, and "traffic police," whose duties included supervising street cleaning. Jewish law, through strictures enjoining individuals to bury their wastes far from people's homes and by calling for streets to be washed daily, gave Jerusalem what must have been pristine sanitation for the second and first centuries B.C.

Athens and Rome both struggled with garbage. The Greek city introduced its first "municipal dumps" around 500 B.C., and also the first known edict against throwing refuse into the streets. Imperial Rome's size led to problems of unprecedented scale, because despite its noted water supply and sewage systems, the city of the Caesars was choking in the wastes generated by its population of more than a million people. Scholars have speculated that the fetid quality of city life may have helped bring about the downfall of Rome by driving the leadership group away to the mountains or the sea. Similarly, in the tenth century, the Japanese moved their capital from Nara to Tokyo because the former imperial city was drowning in its own waste.

In medieval times, waste hardly loomed as a problem for Europe's impoverished, dispersed populations until perhaps the thirteenth century. As people moved to cities, however, often bringing their cattle, poultry, and swine with them, they stumbled toward the reinvention of sanitation, beginning to pave the streets and fitfully to prohibit dumping here and there. By Chaucer's time (the late fourteenth century) when the population of London had climbed to forty thousand, the city's wards were overseen by bailiffs aided by "rakers," who were supposed to gather up refuse and cart it away once a week. But it was not until the Industrial Revolution that Europeans (and Americans) stopped throwing their garbage around pretty much where they pleased.

In the twentieth century, human wastes and trash were firmly swept out of sight, only to reappear at the century's end as a topic of social conversation. Fascinated and appalled, people tried to top each other's incredible (and sometimes apocryphal) unsavory anecdotes. For example, did you know that New York City's Fresh Kills landfill is so high that it will soon block the flight path into Kennedy Airport,* that the disposable diapers used in the United States annually would stretch to the moon and back fourteen times, or that more than five thousand pieces of garbage jettisoned in space during NASA maneuvers currently orbit the earth?

Part of the drama of municipal waste lies in how far things seem to be getting out of control. While landfills overflow, some incinerators shut down for lack of refuse; localities battle over jurisdiction and turn away "foreign" garbage. In the spring of 1991, New Jersey's Essex County incinerator was importing trash from Brooklyn while neighboring Bergen County trucked its wastes to Virginia. On New York's Long Island, four towns only a few miles apart failed to meet state and federal deadlines for closing polluting landfills because they had not

---

*Often cited, even by Vice President Albert Gore, this potent image represents a bit of an exaggeration. The landfill will not in fact interfere with air traffic but will loom larger than any other coastal landmark between Maine and Florida by the time of its closure in 2005.

been able, separately, to plan for affordable, environmentally sound waste disposal—and local politics and institutions precluded their pooling efforts. In Rutland, Vermont, the Vicon Company built and then immediately abandoned an incinerator designed to replace several town landfills, after construction costs had escalated, environmental requirements had mounted, and an ashfill permit was denied. When a somewhat overwrought Vicon executive dumped incinerator ash from a bucket onto the floor at the final town meeting on the project, the symbolism of the gesture could not have reassured citizens present about the future management of their garbage problem.

Another reason that household garbage has emerged from the kitchen into the parlor is that people can visualize their own contribution to the problem in concrete terms. We are able to count the costs much more easily than we can those of most other public services. Further, recycling can allow us to cut those costs. Individual efforts to reduce waste, from carrying a string bag to the grocery store to washing and reusing plastic bags to dissolving packets of soap powder to refilling detergent containers, link closely with collective efforts such as sorting paper and containers to be processed for reuse.

In recent years whisking garbage bags out the back door has become pleasanter, yet doing so produces unease. Our household trash has become lighter and more fragrant as the portion of paper and plastic packages, aluminum cans, envelopes and fliers has grown. The *idea* of garbage, however, weighs more heavily than ever.

Consider, for example, how the garbage crisis has complicated the experience of consuming a TV dinner. Until recently a frozen dinner, packaged in cardboard, swathed in plastic wrap, and dished into a microwavable Styrofoam tray doubling as a dinner plate represented for many Americans the very model of modern sanitary culinary convenience (if not taste). As a *product* it has great attractions: It's bright, unsullied, and neatly apportioned for immediate and individual use. Some of the élan is seeping out of the packaged dining experience, however, as pleasure in the fresh, ingenious wrappings, the ease of popping a full meal into the oven, potholders our only tools, is edged by awareness that the whole thing comes wrapped in three separate layers of packaging. Styrofoam, plastic wrap, and cardboard—the leavings of a rather modest repast—in occupy a fair amount of space in our garbage bag. Won't the plastic survive hundreds of years in a landfill though we only used it for a few minutes; couldn't it be used again or melted down? If it's incinerated, won't the chemicals evaporate into the air or leach from the incinerator ash into underground water supplies? The box seems benign enough, but it too may not biodegrade for eons, and when it does, the chemicals that give the pictured lasagna its red color may seep into groundwater as well.

Thinking about groundwater opens up broader vistas of concern about developmental industrial and agricultural wastes individuals have less power to affect. On a planet whose surface is over 70 percent water, oil, fertilizer, pesticides, and industrial chemicals used on land to produce what we consume inevitably find their way into rivers, lakes, aquifers, and then the sea. The soil itself regulates the chemical composition of the air and water. Exchanging gases with the lower atmosphere and selectively releasing soluble chemicals into surface and underground water, it releases pollutants as well. The transfer of pollutants from atmosphere to water is also impressively efficient, and it can work in reverse too: Some evidence indicates that chemicals may be redistributed from river- and lake-bottom sediments to surface waters and the air.

As we consider how to dispose of consumption and production wastes, both dumping and burning involve tradeoffs among air, land, and water. As we are learning, in the biosphere there's no place to hide. Take the case of the refrigerator. Although its cooling apparatus usually lasts for a very long time, people find many reasons for "junking" one fridge for another: Though a substantial agglomeration of metal, a new fridge is relatively inexpensive to purchase; when the kitchen decor changes, repainting a refrigerator is messy; internal plastic fixtures break off; state-of-the-art ice crushers, water chillers, salad crispers, adjustable wine racks, etc. are not available in older models. If the old refrigerator ends up in the landfill it will displace around 54 cubic feet of space. Refrigerators and other appliances that go to scrap dealers get sliced up into much smaller pieces, some of which are smelted into iron again. Nonetheless, cubic yards of shredded plastic, insulating foam and toxic chemical coolants, paint chips, and rubber tubing are incinerated into a fairly potent ash, or landfilled and washed in liquids that eventually leak from the dump.

Moreover, long before the old refrigerator was unplugged, its fabrication had already created quantities of debris. Extracting the metals for its walls had left piles of iron tailings. Melting and melding its materials precipitated metal and plastic residues onto the factory floor and released chemicals and gases into the air. During both manufacture and disposal chlorofluorocarbons (CFCs) wafted into the ozone from coolants flowing through its channels and ducts.

Whatever the source of their particular concerns, over the past few years Americans have spent inordinate amounts of civic energy on the household garbage problem. In part they are spurred by mounting costs, and in part, as we have seen, they are responding to a problem they are relatively able to affect. Since the early 1980s, they have also been able to go after industrial polluters.

Together, garbage politics and the local hazards of industrial waste have provided major spurs to a new populist environmentalism. Although the initial phase of environmental activism in the 1960s was spurred by a scientist, Rachel Carson, the first big impetus for the next wave, at the end of the 1970s, came from a housewife, Lois Gibbs. More than any earlier American environmental activism, the surge that started in the 1980s came from the grass roots. Driven by resentments at costs and damages inflicted from above, the movement gained impetus from Gibbs's facedown of New York State and the federal government over toxic waste in her Niagara, New York, neighborhood, an area that became notorious under the moniker "Love Canal."

In 1978, Lois Gibbs galvanized her neighbors, whose homes were constructed on top of the Love Canal toxic waste dump, to fight. Gibbs made an unlikely giant killer. A high school graduate married to a chemical worker, she lived in what she said looked like "a typical American small town that you would see in a TV movie." She was oblivious to the problems of toxic chemicals until she read in the Niagara Falls *Gazette* in June 1978 that her six-year-old son's elementary school was built directly on top of

a dump where Hooker Chemical and the U.S. Army had jettisoned toxics for twenty-three years. When school authorities refused to deal with her concern about connections between the
toxic chemicals and her son's epileptic seizures and low white-
cell count, her door-to-door quest for signatures on a protest
petition exploded into a national controversy.

During the summer of 1978, her neighborhood crusade became an epic struggle, eventually escalating from meetings with
the New York State commissioner of public health to meetings
with New York's governor Hugh Carey to scientific debates over
the spreading illness and genetic disorders among neighborhood residents, to congressional testimony, to a symposium at
the White House and a Niagara Falls ceremony at which President Jimmy Carter contributed $15 million in federal funds to
pay for relocation of Love Canal area residents. Gibbs's saga
made a gripping drama, with a heroine suddenly catapulted
from the kitchen to the public arena, one moment debating the
governor and the next, breaking down in exhausted tears after
forgetting her son's sixth birthday and absentmindedly devour-
ing most of his cake. The central conflict ranged the local
neighborhood people, impelled by fears for their children and
unborn infants, against the city and state governments, which
seemed determined mainly to safeguard Niagara Falls's tourist
appeal and stave off liability claims. By the time the powers that
be had cried uncle, Gibbs and her cohorts had learned how to
run a political movement.

Aided by sympathetic scientists, they undertook their own research. Their findings gave evidence of a pattern of disease cor-
responding to underground streambeds, a 50 percent to 75
percent chance of miscarriage, a 56 percent chance of birth defects, and a significant incidence of central nervous system diseases, hyperactivity, urinary disorders, and respiratory problems
in the community as a whole. Working to make their case, they
learned that they could catch their adversaries' attention on the
evening news.* The publicity gained valuable allies, including

---

*Press coverage of the Love Canal Homeowners Association studies on
neighborhood illness forced reactive research on the part of government departments. The state's findings consistently weakened the alleged cause-and-
effect relationship between the Hooker dump and endemic community illness.

state legislators, local congressmen, New York's U.S. senator Daniel Patrick Moynihan, and even Jane Fonda. After her neighbors were relocated, Lois Gibbs went on to found a nationwide organization, the Citizens' Clearing House on Hazardous Waste, to pool grass-roots energies against toxic pollution.

The gravity of the health hazards that galvanized the movement are still under dispute. In the decade after the relocation, the scientific community came to characterize the crisis over Love Canal as an overreaction to exaggerated public fears. In 1991, as the city put the vacated houses around the evacuated dump site up for sale, experts were still debating the actual effects on health of the undoubtedly toxic chemicals. Like numerous other assessments of chemical hazards, the accuracy of the allegations by scientists working for the Love Canal Homeowners versus those of the state's epidemiologists has not been determined with certitude.

Nonetheless, Love Canal became a beacon for grass-roots groups mobilizing to combat environmental risk. Although the word "conservation" was seldom heard, the idea of preventing pollution by saving resources also began to play a role in grass-roots action. During the 1980s a new politics for management of household trash emerged, one that was dominated by local opposition to incineration and support for reprocessing materials in garbage, or recycling, and waste minimization (also called "source reduction"). According to the waste management industry's trade organization, the National Solid Waste Management Association, by 1991, 86 percent of Americans felt that communities should not consider building new incinerators until they had established strong recycling programs—a political sentiment incinerator vendors could not openly oppose.

However, this new sort of environmental thrift—which also influences the government policy for industry—has very shaky economic underpinnings. Over the past few years, federal and state antipollution programs have begun to pressure industry to use resources more carefully. Both forms of conservation, in the household and in business, have to fight the system of incentives

---

Dr. Huffaker, the state's site safety officer, admitted to residents that "it will smell like hell. It will smell like Hooker. But," he declared, "it won't hurt you." (Eventually workers on the canal site were required to wear protective boots, gloves, and helmets, some even gas masks.)

embedded in our economy. Despite good intentions and a significant shift in rhetorical emphasis, rewards for production and waste are still great while the rewards for reducing use of virgin materials, reusing products, and recycling materials are minimal.

On household garbage, the idea is that recycling should pay for itself through the sale of what is collected. More important, the sale of materials should prove that it is cost-effective to recycle. Thus it is assumed that the market itself will provide the ultimate test of viability.

Unfortunately, the market will not carry recycling while our economic incentives overwhelmingly reward exploitation and extraction of virgin materials. Only a tiny fraction of plastics are recycled and the use of plastic materials is mushrooming. While people are depositing their newspapers, bottles, cans, and plastics by the curb, too much of the packaging is piling up in warehouses waiting for buyers. Those who love the thriftiness of collecting and returning waste could lose heart when they find out that what they saved was eventually dumped in the landfill after running up storage fees. If recycling and source reduction periodically flunk the market test, waste reduction may come to look "unrealistic."

Even now, businesses have somewhat more compelling economic reasons to curb consumption of materials—but not compelling enough. In the long run, enterprises can save money on both procurement and waste disposal by more efficient use of resources. Yet at present the private sector appears to be investing as much in fighting the regulations that would force waste reduction as in retooling processes for conservation.

In view of these realities, tinkering with the system will not get us very far in handling problems of this magnitude. To deal adequately with our waste crisis we will have to rework policies directly affecting the U.S. economy as a whole. American political values, however, place peculiar limitations on our ability to solve systemwide problems. We actively resist ceding to government power over our economic lives.

Given the enormity of the crisis that we face, we do not have time to revamp our political culture in order to deal with it—

even if we wanted to. Thus, this book will explore solutions consonant with our political values that will enable us to bring about the major systemic changes we need. We will consider how to use our strongest suit—entrepreneurship organized by the market—to curb our use and waste of resources while creating a new sort of plenty.

The book aims to supply all the building blocks essential to understanding how our society can come to grips with a problem whose sources lie deep within our way of life. Starting with our forebears, the incredible natural endowment of this continent and their struggle to seize control of it from the King of England imbued them with both egalitarian individualism and a strong conviction that the market—not government—was the best arbiter for people's affairs. And the outpouring of entrepreneurial invention during the nineteenth and twentieth centuries appeared to prove them right. With the dramatic changes after World War II, however, the story grows much more complex—and threatening.

At least since the later 1960s, Americans have been trying to deal with the material fallout from the good life. For the most part, however, our successes in stemming the flows of pollution have been dwarfed by the vast stream of production and consumption rolling steadily onward. When one looks at all the sources of polluting residues—garbage, sewage, industrial emissions, agricultural runoffs, and nuclear debris—it becomes clear that current forms of waste management, including recycling, will not solve our problem. Very simply, our best means of dealing with the overall waste crisis lies in reducing our use of materials.

Though we need them all, none of the instruments usually advocated to reduce waste—neither technology nor private entrepreneurship nor government regulation, nor all three together—is adequate to the vastness of the problem. Cutting way back will require the exertion of extraordinary pressure to change the way the system works now. Ultimately, we have to find a way to make our most efficient organizing mechanism, the market, work for conservation rather than exploitation. Within the current American political context, that will be no easy task, for it will require us to enact government policies that will change the way production and consumption processes work.

Clearly, restructuring our economic order to salvage the environment will involve major shifts in the way we think about property, production, and progress. Paradoxically, if we want to use the market effectively, we can no longer afford to consider government an alien force. Only if we acknowledge government as our agent in changing incentives for the private sector can we hope to make the most of our entrepreneurial strengths.

In defining garbage as everything flowing into the land or water that man-made and natural systems cannot handle, my description of how wastes are changing our environment for the most part excludes air pollution. Yet the distinction between air pollution and other kinds of waste cannot be strictly maintained. We cannot, in my view, hope to deal with our waste crisis without changes so fundamental that they will work for the environment as a whole. Though implementing these changes may alter our pursuit of material happiness somewhat, it will not seriously obstruct our quest for the good life. Unless we make conservation the cornerstone of future economic progress, our dream of plenty and a good life will not survive.

2

~~~~~~~~~~~

Freedom and Wealth in
the Minds of Americans

*Every individual ... by pursuing his own interest ... fre-
quently promotes that of society more effectually than when
he really intends to promote it.*

—ADAM SMITH, *The Wealth of Nations*

*It isn't the government's job to tell people what kind of air
they should breathe. Each person should decide for himself.*

—ART BUCHWALD, *Sierra* (November–December 1981)

FOR AMERICANS, PHYSICAL limits are merely temporary
obstacles—like the Rocky Mountains to the westward-bound
pioneers. If you keep moving, anything is possible. As for imped-
iments less imposing than a mountain: Mow them down. Contending in what philosopher William James called "the
immemorial human warfare against nature," Americans grew
confident of their advantage. Like the saga of the pioneers, clearing away the forests for farms, pushing rails across the prairie,
changing the courses of rivers to make the desert bloom, many
of our traditional images of national grandeur depict a path to
abundance through physical exploitation.

Because our confidence in a limitless tomorrow has been prof-
ligately rewarded, it has shaped the way we think, not only about
nature but about society. Our nineteenth-century surge across
the continent led Americans to riches our forebears could hardly
have imagined. With so much there for the taking, each man
could be expected to make good on his own if he was willing to
go after it. "I contain multitudes," declared the poet Walt Whit-

man, expressing a sense of unbounded personal reach shared by many nineteenth-century Americans.

Thus we prefer to view our conquest of a continent, despite its massive scale, as the work of heroic individuals, solitary souls with great ideas. While we were pushing back the frontier, popular heroes were backwoods fighters like Davy Crockett or the sharpshooting sheriff Wyatt Earp, and occasionally the outlaws they pursued. In the later industrial boom, people's imaginations were most stirred by the exploits of inventors like Thomas Edison, pictured working alone in his laboratory, and of the great corporate leaders like Andrew Carnegie and John D. Rockefeller, who held entire industries in their hands. Although we revere the revolutionary statesmen and lawgivers who were our Founding Fathers, for the past century American role models have not usually been public figures but rather "self-made men," wresting from nature and their fellow citizens whatever share of wealth they could commandeer.

More than a century after Whitman, and after the American Revolution he celebrated, extraordinary changes in our material lives call into question the applicability of traditional political attitudes. Physical limits press in on us from all sides. The Japanese and Europeans have been gaining on us economically, and the oceans between them and us seem smaller. Harder still, our land itself appears to resist our absolute claims on it. If our economic growth is making our habitat unlivable, all our individual victories will not stave off defeat in the war. Although Americans certainly command the wealth and skills to engineer a different, more respectful, relationship with our still extraordinary resource endowment, to make that shift we will have to change our images of abundance while we alter the processes for creating it. We will have to amend beliefs we have shared as a people since the eighteenth century of how we can best foster "the pursuit of happiness."

Though "ideology" is for most Americans a term with negative connotations, we are, in fact, the last ideologues of the twentieth century. Now that Marxism has fallen victim to its own internal contradictions, the United States represents the final bastion of true belief in the other great ideology that developed in tandem with industrialization. While many of our fellow industrialized nations have economic and political institutions similar to our own, Americans hew to a set of ideas that we call "capitalism" with unique single-mindedness.

It is beyond the scope of this inquiry to consider in great detail the character of our mind-set, or how it evolved; this question has been richly explored elsewhere. However, our view of ourselves as capitalists is key to coping with the waste that has resulted from the way we make things and how we think about the things we've made. And we will touch upon it briefly.

The way we think about economic and political life results from a synergy between our geography and our history in which causes are intertwined and difficult to sort out. Our revolt against an alien government, the exceptional equality of condition among the early settlers, the expanse of our territory, our openness to immigration and migration within our borders, the land's wealth—together these provided hothouse conditions for the fruition of certain seventeenth- and eighteenth-century notions of human nature and society.

From the beginning, the fortuitous conjunction of ideas and circumstances fostered a mind-set that has made it easy to grow but harder to govern. Our forebears' access to abundant land and resources encouraged an individualism that is unique in the Western world. American success in applying the ideas of Enlightenment philosophers—particularly about individual rights as grounded in private property—would probably have amazed those philosophers. In this peculiarly fortunate situation, however, these ideas have bred habits of thought and action on public policy that often obstruct effective governmental action today.

The same conditions did of course foster the formation of private organizations wherein individuals could work together to further their goals. Writing in 1834, Alexis de Tocqueville described the American propensity he saw then for resorting to cooperative associations whenever faced with a new problem:

If some obstacle blocks the public road halting the circulation of traffic, the neighbors at once form a deliberative body; this improvised assembly produces an executive authority which remedies the trouble before anyone has thought of the possibility of some previously constructed authority beyond that of those concerned.

This characteristic American response has provided the impetus for political movements that have changed American life. Our national penchant for joining in extragovernmental groups to further particular causes stretches back to the Revolutionary Committees of Correspondence and the vigilante posses of pioneer days. During the nineteenth and twentieth centuries a combination of moral fervor and grass-roots organization drove the nation toward the abolition of slavery, prohibition, municipal reform, and women's suffrage. While moral questions (like slavery, temperance, and abortion) have had considerable power to galvanize local political energies, so has shared vigilance over private property. The modus operandi of many grass-roots environmental crusades—"Not in My Backyard!" or NIMBY—expresses very well the roots from which they spring, in which the "personal" and "property" stakes figure prominently. In the case of Love Canal, as with analogous movements, the bounds of private property are extended to include commonly used local turf—the whole neighborhood. Issues of individual health and safety and municipal costs interwine closely with property values, as people focus on protecting themselves "where they live."

For the last three decades, as we have seen, grass-roots activism has provided the motive power for U.S. environmental progress. The impetus of this grass-roots activism is essential if we are to change our lives enough to salvage our habitat—yet it is not sufficient. Having arrived at what is a global physical impasse, all the best efforts of individuals to save and conserve cannot change the larger system quickly enough. For our effort to match the magnitude of the challenge, we need government to redirect incentives for production as well as consumption.

Most Americans, however, do not see it this way. Thomas Jefferson summed up American attitudes toward government prevailing in the eighteenth century when he said in his first inaugural address:

> With all these blessings, what more is necessary to make us happy and prosperous people? Still one thing more, fellow-citizens,—a wise and frugal government, which shall restrain men from injuring one another, which shall leave them otherwise free to regulate their own pursuits of industry and improvement, and

shall not take from the mouth of labor the bread it has earned. This is the sum of good government, and this is necessary to close the circle of our felicities.

To a surprising extent, despite two subsequent centuries of growing government involvement in all aspects of Americans' social welfare, Jefferson's minimalist view of government has continued to reflect the ideal still cherished by many.

Unlike the Europeans we have lacked a strong national tradition about leaders', or the state's, responsibility for the people's general welfare. In our case, we skipped the feudal period, with its sense of strong reciprocal obligations, and our early settlers lived within the ambiguous power relations of colonialism. Our earliest experience, for roughly a century and a half of colonial rule, featured arms-length administration by various agents of the Crown and increasing assertion of control by local citizens. In our first taste of government, our rulers actually tried to stifle private enterprise on this side of the Atlantic in order to prevent colonial industry and commerce from competing with that of the mother country. Our revolution against royal and parliamentary restraints on the productivity of American farmers and burghers, and in favor of local citizens' control, set an enduring tone for relations between public and private sectors.

Further, our political tradition depends heavily on the ideas set forth by our Founding Fathers and in turn on the Enlightenment philosophers, especially John Locke, who profoundly influenced them. The thinking of these philosophers and of the economists of the same period lays significant claim upon us still. What we have celebrated with unique tenacity from the beginning of the American experience is the right of the individual to pursue opportunity free of government exactions, a right secured by ownership of private property.

The earliest image of the independent American was the subsistence yeoman farmer, whom Thomas Jefferson considered to most closely approach the ideal of self-sufficient existence and who exemplified individualism to Jefferson's countrymen as well. In their fields, barns, and kitchens, with their wives and children, rural Americans produced most of what they consumed. The material conditions of American life changed greatly, of course, in the century after Jefferson's address, as industry beckoned and, at length, more and more farmers left the land for the city. Even when the factory system transformed the self-sufficient yeomen into cogs in a larger machine, however, the individualistic ideal retained its potency.

From the outset, America's enterprise depended on the land's extraordinary abundance. Forests were for many of the early colonists and frontiersmen obstacles in the way of farming—clearing them made fuel wood so plentiful that even where possessions were few, fires of manorial splendor often blazed through much of the year. Later, the European calculus of cheap labor and dear energy was reversed here as factories designed to run on waterpower sprang up along rivers and streams all over New England and down the Eastern seaboard. Much later, geysers of oil literally flooded the southwestern plains—and the market. With such a lucky lien on nature's bounty, who would count the costs?

The same sense of largesse pervaded Americans' relationship with the land. Although thrift was seen to be a quintessentially American trait among the more settled areas of the East, the prodigal spirit of the westward surge came to dominate our idea of ourselves. Stories of the dramatic tug-of-war with the Indians over the Great Plains provided endless episodes for a serial national epic, in which the massive buffalo hunt that finally wiped out herds of millions provided a symbolic parallel with the removal of the native tribes. In the century during which we spread across North America, the extraordinary expanse of land at our disposal gave reality to the idea that each person could be economically self-sufficient. So few people, so much real estate! For much of the country, Washington seemed almost as far away as Windsor Castle had to the colonialists. Though the government kept order and occasionally supported private efforts (most notably the railroads) that would foster general economic expansion substantially, people had enormous elbow room in their individual pursuit of wealth and, presumably, happiness.

If a homestead was used up and failed to yield, one could move on to another plot, perhaps in another territory. Even towns were disposable, built up and then left behind as mining or commerce moved on. Although we imagine our ghost towns all to be in the wild West, in Iowa, for example, more than 2,200 towns and post offices were abandoned between 1838, when Iowa became a territory, and 1930. Kansas counted more than 2,500 "geographically extinct locations" between 1852, the year the area gained territorial status, and 1912.

After the frontier officially closed, America's internal migrations did not cease, but the promise of wealth emanated increasingly from technology as much as real estate. American values strongly supported the quest for industrial progress. As the factory system expanded, growing differences in wealth occasioned surprisingly little envy. For Americans, social mobility was expected to come from growth, not redistribution. Money was closely linked to virtue both as an incentive for industriousness and as its reward. Early in the century Alexis de Tocqueville had marveled at the "avidity with which the American rushes forward to secure this immense booty" of "a boundless continent." Wealth conferred so many good things in addition to comfort— status, virtue, honor. Its acquisition has been our highest form of endeavor. My time will come.

The Pennsylvanian Albert Gallatin, who viewed his countrymen from the vantage point of one reared and educated in Europe, observed in 1808, "The energy of this nation is not to be controlled; it is at present exclusively applied to the acquisition of wealth and to improvements of tremendous magnitude." Thirty years later, an Austrian observer, Francis J. Grund, echoed Gallatin and Tocqueville when he noted the intense focus of Americans on manufacturing: "Business is the very soul of an American. . . . It is as if all America were one gigantic workshop."

With the later growth of commerce and then manufacturing, the theories of Adam Smith and David Ricardo about the benefits of markets freed from government interference took root in fertile ground. Thus the American economic experience in the nineteenth century gave further credence to the idea that if you set people free from government interference and let the "invisible hand" of the market allocate resources for investment and consumption, people will prosper and so will their nation. From around 1800 onward, the U.S. economy grew faster than any other, and at the end of the nineteenth century we passed Great Britain to become the world's leading manufacturer. Moreover, for much of their history, Americans had lived better than most of their British and European cousins, enjoying generally higher wages as well as the psychic and physical pleasures of all that space.

In the second half of the nineteenth century, the ideas of biologist Charles Darwin and philosopher Herbert Spencer reinforced existing American predispositions about how society should work. With their depiction of biological and social evolution through fierce combat they provided an ennobling rationale for bloody struggle in a competitive marketplace. Given the unparalleled opportunities Americans enjoyed, the driving power of competition, prescribed by Ricardo and Smith, as well as Darwin and Spencer, appeared relatively benign. Perhaps for this reason, Americans embraced Darwinian social and economic attitudes even more readily than their British counterparts, who were running short of land and were conscious much earlier of the harsh consequences of the Industrial Revolution for workers' lives. To most Americans the scramble seemed to work well: Everyone had the same opportunity and it was their own fault if they didn't take advantage of it.

The idea that maximum competition kept everybody honest and efficient bred a preference for conflict as the chief arbiter of differences. Americans tended to resist the idea that differences might better be ironed out prior to a shootout or a showdown. Enormously rich tycoons like John D. Rockefeller and J. P. Morgan, who strode forth into our consciousness in the latter part of the nineteenth century, seemed proof that if you let nature take its course through the "natural" arbiter, the marketplace, the fittest would survive and thrive. Their achievements reinforced the notion that society, and by that we meant government, should not interfere with an ultimately beneficent process.

Extending the concept still further, the public sphere itself could be seen as a sort of marketplace. In the political realm as in the world of commerce, the aggregate of people's decisions about what they wanted should strike the right balance about what was produced: Citizens could decide through their votes whether given public services were worth enough to warrant paying taxes. By and large, however, the fact that choice could be so much more clearly and immediately reflected in the actual marketplace than in any electoral process enormously enhanced the legitimacy of decision-making and resource allocation through the private sector rather than through government. The marketplace was seen as a democratic process while the economic machinations even of a freely elected government smacked of autocracy and deadening inefficiency.

As we have seen, Americans adapted extremely well to the changing marketplace. In the decades following the Revolutionary War, scientific education in the United States lagged woefully behind that on the continent, but that gap probably didn't matter very much to our national progress. Technology was still ahead of science in mastering methods before scientists really understood the principles on which they were based, methods that in turn often opened up new areas to scientific investigation. And Americans proved to be particularly gifted at technological innovation. As a society we have long prided ourselves on our "ingenuity," our down-to-earth native cleverness and ability to solve problems large or small, using whatever means are at hand. Although to a great extent Americans in the post-Revolutionary era lacked the specialized technical training that European artisans acquired through apprenticeships, they were able to improvise freely. As they made a virtue of necessity, "jack of all trades" became a ubiquitous American term with generally positive connotations.

Social adaptability reinforced technical flexibility. In Europe new machines and the prospect of cheaper goods threatened the economic and social security of entrenched classes and guilds. In England the momentum of innovation eventually slowed as the bourgeois businessmen who had driven it aspired to take their winnings and join the aristocracy. But in America no important interests were menaced and no aristocracy existed to be joined. At the end of the nineteenth century, America stood poised on the brink of a second Industrial Revolution—in production processes and technologies—that in material terms, and social consequences, would dwarf the first. With our strong faith in individual economic effort and hope for wealth, Americans responded enthusiastically to the renewed promise of abundance.

And we were commensurately rewarded. We produced goods so efficiently that they began to stream from U.S. factories like oil from a gusher. Wages rose, though somewhat unevenly and insecurely, given the frequency of business "crashes" during the nineteenth and early twentieth centuries—but it was the extraordinary aggregate *output* of industry that changed people's lives. Ordinary individuals gained access to comforts and conveniences that even wealthy households could not have enjoyed a century before. Henry Ford made the automobile an affordable possibility for every family. Music and motion enlivened and accelerated daily perambulations, as first the automobile, then the radio, then the automobile *with* the radio became nearly universal fixtures in the space of a few short decades.

At the turn of the century, Americans began a romance with technological progress. The possibilities for "progress," meaning the improvement of material life through science and technology, seemed endless. Underpinned by a public school system that was then the best in the world, graduate education in science took off in the United States after the Civil War. Corporations like General Electric, Standard Oil, and U.S. Steel began to hire scientists, and many of the newly emerging managerial elites had themselves been trained as engineers.

Throughout the first decades of the twentieth century, Americans enshrined the engineer as cultural hero combining the promise of science with pragmatism. More than anyone else, the engineer embodied the human ability to triumph over nature. As literary historian Cecilia Tichi points out, the engineer appeared as the hero of more than a hundred silent movies and the hero in best-selling novels that sold over five million copies between 1897 and 1920. Humorist H. L. Mencken satirized the way in which popular regard for the profession led those in lesser fields to appropriate some of its gloss, as, for example, the bedding manufacturer "who first became a *mattress-engineer* and then promoted himself to the lofty dignity of *sleep-engineer*," or the "beautician who burst out as an *appearance-engineer,* and the *demolition-engineers* who were once content to be house-wreckers."

General Electric's Charles P. Steinmetz, an electrical engineer of genius, was just such an icon in real life: a leading researcher for G.E. from 1893 to 1923 who became a public figure. A political philosopher who argued on the grounds of efficiency that government should be run like a corporation, he was a darling of the media and a draft candidate for the position of New York's state engineer. For the American public Steinmetz symbolized the era's enthusiastic embrace of technical achievement and productive efficiency under the corporate aegis, a passion that would endure among average Americans until the Great Depression and the presidency of engineer Herbert Hoover.

Steinmetz's devotion to productive efficiency was widely shared by his contemporaries, the most influential of whom was Frederick Winslow Taylor, who redesigned work according to principles of "scientific management," also known as "Taylorism." In industry, Taylor's dictum that there is one right way to accomplish a task led employers to break down the work process into numerous separate segments and standardize repetitive patterns in an effort to achieve total control over workers' sequence of motions. Workers acting as cogs in a machine could be expected to achieve machinelike productivity. Henry Ford's pathbreaking assembly line, combining continuous flow with interchangeable parts in a careful calibration of time and motion, enormously increased productive output.

Toward the end of the nineteenth century, the setting for family life was also redesigned to combine, in the suburbs, rational systems and rural individualism. The new middle-class sanctuary, which would become the quintessential residential model for Americans, originated with a group of prosperous London Evangelicals early in the century, when Great Britain was at the apogee of her industrial momentum and urban squalor. The garden cities movement took hold on this side of the Atlantic somewhat later, following our own mid-century industrial boom and the expansion of the American professional and managerial class. With the spread of the railroad and the trolley, affluent families were able to seek the American dream outside but near urban areas.

The idea of the suburb was to remove one's family from the

dirt and commerce of the city to a harmonious rural enclave. As historian Mary Corbin Sies has pointed out, for many young professionals and business executives in the 1870s and '80s, the "social correctness" of suburban life became an ideology. Though clearly interdependent with the nearby city, suburban existence catered to the idea of independence and local control. As she states it, "The controlling idea of the proper residential environment was one in which every family resided in a one-family home with plenty of yard within a locally controlled, homogeneous community." Because it became the American ideal, the consequences of that belief have been far-reaching.

Then as now, the attraction of these new residential enclaves flowed from the traditional American glorification of self-sufficient rural life and rejection of the city as unhealthy and vaguely immoral. Set apart from the metropolis, with careful (sometimes lavish) spacing between dwellings, the suburb reinforced the idea of separateness that has always been powerful for Americans: between individuals, and between public and private realms. As social historian Lewis Mumford put it, the suburban movement was an attempt to make one family the community. Retreating from commercial and civic affairs into securely private property, each family had space to create its own world.

Full of faith in science and doubt about society, the suburban pioneers, like numerous Americans before them, aspired to create a new and perfect environment. In communities like Short Hills, New Jersey, founded just eighteen miles by rail from New York City in 1877, residents called on experts to create a setting conducive to health, virtue, rationality, and social order. They saw the countryside itself and the beauty of nature as a major source of spiritual uplift. But they also viewed technologies for sanitation, heating, and lighting, as well as the design of housing, roads, neighborhoods, and schools, as key. Amid modern conveniences they aimed to shape social harmony for a homogeneous group of like-minded citizens and create an "ideal home environment" for each family. Sanitary engineering would preserve suburban yards as bucolic patches of "nature" insulated from the unsavory realities of human waste disposal. Short Hills constructed a complete waterworks, paved roadways, gas lighting (in 1880), a sewerage system (1888), an electrical system (1895), and fire protection with a fifty–call box alarm system (1895).

Imbued by the ethos of production efficiency, the suburbs also sparked a consumption binge. Cultural life in the new communities reflected the prevailing faith in "the engineering approach" to the management of people's affairs. As the only daytime adult inhabitants of the suburbs, women were essential to the smooth functioning of the total domestic unit that was the commuters' village. In the parlance of the day, women had become "rational managers" of the home. Within an atmosphere dominated by the advances of science—with debates raging about sanitation and disease prevention, and popular fascination with the new mechanical appliances—women were urged to act as "domestic technicians." To guard the health of their families and learn to use their new tools efficiently, they were advised to seek the advice of "experts." In their condition of relative isolation, shopping to equip their families for the ideal material life became part of a woman's vocation. As the suburbs expanded at the turn of the century and again in waves after both world wars, housewives' conscientious provisioning of their homes spurred the consumption that kept factories humming.

From the start of the twentieth century, the automobile appropriated the symbolic and real center of this expanding material existence. Finally fulfilling dreams of mobility that had driven Americans since the days of stagecoaches and covered wagons, the motor car made possible virtually universal suburbanization. The auto also seemed peculiarly American in reinforcing aspirations toward individual control. Unlike other more sociable modes of transport, each vehicle, enclosed and self-sufficient, forged its separate way down the road.

Because the automotive industry provided the lynchpin for the emerging U.S. industrial economy, it is not surprising that the most important battle for the minds of American consumers was fought by its earliest titans. Though himself an economic revolutionary, Henry Ford's impact grew out of his genius for machinery, for process, and for finance. Although he believed deeply in progress, Ford's personal aversion to excess, waste, and indulgence disposed him to ignore the briberies and seductions involved with marketing and sales. Simple in tastes and narrow in perspective, he worked to perfect a utilitarian and *cheap* car for all Americans. When he had done so, he thought the all-black Model T should be good enough for just about anyone and he kept on producing it from 1908 until 1927.

But in the next stage of the struggle, the great innovator of mass production—surprisingly—signified the thrifty past and Alfred Sloane, Jr., the president of General Motors, represented the prodigality ahead. Sloane was a man of the future. He well understood that building a better mousetrap—even a cheaper one—was only the first step. Like credit flows, the customer himself had to be *managed*. Inspired by the profits reaped in the world of fashion by constantly changing styles, he devised the basic marketing innovation underlying our present patterns of consumption—and waste—"the annual model change." Given the capabilities of the mass production economy, the marketing emphasis was veering radically from satisfying needs to fulfilling desires. Underlying the sales pitch lay a profounder imperative: To keep the game going, consumer attitudes had to be shifted from buying for necessity to the necessity of buying. The leading Muncie, Indiana, newspaper put it very directly in a 1925 editorial: "The way to make business boom is to buy."

These changes did not come without a struggle. Many of Ford's contemporaries shared his apprehensions, as shown by a compilation of quotes from popular and professional journals between 1926 and 1928 inveighing against consumer credit: "enervating to character because it leads straight to serfdom"; "tending to weaken the moral fiber of the Nation"; and "breaking down character and resistance to temptations, to extravagance and to living beyond one's means." But attitudes had been shifting swiftly since the First World War. Writing in 1931, a social critic decried the major retreat from old values during the 1920s, when, as he said, people had started to find it "old-fashioned to limit their purchases to the amount of their cash balances."

Even by the end of the nineteenth century the managerial techniques and ethos of the new corporate organizations had begun to color how people thought about their personal lives. Working in factories, more and more people were forced to live according to the precise rhythms of the clock. Consequently the values of rational planning, efficiency, and expertise gained power steadily. Numerous social critics have since noted how an increased focus on comforts and possessions substituted for the lessening satisfaction in work of factory employees on the assembly lines. Thus a growing array of comforts smoothed the transition from the production-oriented society of the small entrepreneur to the consumption-oriented society dominated by the bureaucratic corporation. On one level, the compensatory materialism grew naturally out of the traditionally high regard for wealth that Tocqueville described as a fundamental trait of "men living in democratic times," most of whose "passions either end in the love of riches or proceed from it." On another level, however, the material revolution had produced a profound transformation in values. As people became increasingly interdependent, they took refuge in the idea of psychic as well as material self-fulfillment as a way to sustain individuality. Already under siege, the traditional virtues of thrift and self-sacrifice were further eroded by the new imperative to consume.

At the same time the business of production took on aspects of the sacred. Our most quotable of presidents, Calvin Coolidge, showed how far the goods and glories of *production* were reshaping popular attitudes when he remarked in 1924, "A man who builds a factory builds a temple," and "a man who works there worships there." The identification of business and religion again provided the central metaphor in a 1925–26 book by adman Bruce Barton (founder of the Batten, Barton, Durstan and Osborne Advertising Agency). Entitled *The Man Nobody Knows,* it depicted Jesus Christ as a great business executive who "picked up twelve men from the bottom ranks of business and forged them into an organization." Barton's Christ was preeminently a salesman.

The link between increasing production and making people's lives better lay in making consumers want what was produced. Therein lay a convergence of personal and public goods: People could achieve "fulfillment" and the economic machine would keep humming. Merchants have always tried to sell things, the more the better. But before the revolution in production, limitations on supply (and correspondingly high prices) effectively ruled the market for manufactures. In the late nineteenth and early twentieth centuries all that changed. When new power sources, machinery, and techniques of mass production brought goods surging into stores, catalogs, dealerships—their manufacturers *needed* buyers. Suddenly the willingness of people to spend became a primary index of the economy's well-being.

In a peculiar twist on the old idea, demand had to satisfy the needs of suppliers. Or, as social critic Vance Packard put it in his 1966 book *The Waste Makers*, "The way to end glut was to produce gluttons." For consumers, it became a matter of course that new things would displace older ones: You had to make room. In a nation that had never confronted serious raw materials shortages, and during a period when real constraints on the limits of our environment to handle our wastes were unimaginable, durability and thrift loomed chiefly as matters of taste and perhaps a dated morality.

A renowned array of popular thinkers certainly thought so. Luminaries including churchmen, social scientists, journalists, corporate executives, and politicians cohered around a new "philosophy of life" that seemed to answer the needs of their maturing capitalist society. This set of values, dubbed the gospel of "therapeutic release" by sociologists Richard Wightman Fox and T. J. Lears, trumpeted the virtues of consumption as a healthy quest for "self-fulfillment."

Highly influential in the twenties and thereafter, the new approach replaced what Fox and Lears characterized as the traditional ethos of "perpetual worth, compulsive saving, civic responsibility and rigid morality of self-denial" with the new values of "periodic leisure, compulsive spending" and a "morality of individual fulfillment."

Founded during the 1920s to propagate the new faith on a popular level, the Coué Institutes were the forebear of numerous self-actualization movements in generations to come, such as Norman Vincent Peale's "power of positive thinking" of the 1950s and Werner Erhard's *est* training seminars of the 1970s. Coué emphasized the control an individual could wield through willing: "Day by day, in every way, I am getting better and better." Applying the same message to the larger picture, the Muncie newspaper advised its readers that "there can be no greater single contribution to the welfare of the nation than the spirit of hopefulness."

A new set of "experts," the social scientists, also weighed in, both with advice on how society should work and with novel tools to assist businesses in understanding and shaping the desires of consumers through market research and advertising. Advertising, of course, was to play the key role in ensuring that consumers consumed. Enthusiasm for advertising went hand in hand with veneration of business and the progress it was said to boost. Sociologist Stuart Ewen has pointed out how far that went: In the aftermath of the 1917 Russian Revolution, prominent voices in the American business community, equating consumption with political freedom, touted advertising as the answer to Bolshevism.

By the 1920s, all sense of limits on industrial production apparently had vanished (as subsequent chapters will detail). By that time the ideological priming of the consumer society had been liberally applied to the sturdy superstructure of American capitalism. Thereafter, despite a significant hiatus during the Great Depression, the sky was the limit. On the supply side, our creative powers seemed to rival those of nature, allowing us not only to manufacture goods so fast they seemed to "grow" but also to invent new combinations of basic elements and then entirely new materials.

For the period of one long lifetime—seven or eight decades—we were able to enjoy unbounded material progress in relatively happy innocence. Not until we were deep into the boom years following World War II did we begin to suspect that our inventiveness might greatly exceed our understanding of what we had wrought. In this century of technological overdrive we had grown accustomed to relying on more magic as an antidote to spells and potions gone awry. American confidence in our forward march through human ingenuity played a key part in bolstering the optimism we depended on, which increased our reluctance to admit to doubt. Though it sustained some hits, our belief in progress through technology outlasted faith in the inevitability of spiritual or societal improvement. It was not until the 1960s that most Americans would worry at all about the environment— and thereafter only in episodic seizures.

As we move through this century's last decade, however, we have come to another major crossroads in our material passage. Because we can no longer avoid facing absolute material limits, our situation represents a profound break with our past. Thus the present juncture will be particularly difficult for us to navigate.

Paradoxically, the magnitude of our present crisis may abet efforts to cope with it—if we acknowledge the gravity of our position. Only in times of crisis have Americans generally been able to transcend ideological constraints on societal growth. At present, our aversion to government deprives us of useful instruments for resetting our course—yet we need the overall guidance of government in managing the transition to a new material order combining environmental and economic efficiency. Throughout our history, economic crises have allowed sharp breaks with the prevailing practice of laissez-faire. At these times, with our ideology in temporary abeyance, we have employed activist public responses in effecting far-reaching institutional change.

The first such major watershed occurred at the end of the last century, when the wastes from the Industrial Revolution threatened to make American cities unlivable. The Industrial Revolution had done much to fulfill our dream, but the communal crises it brought—of urban squalor, poverty, and water and air pollution—would not yield to energetic individualism or to commerce.* The first American garbage crisis, which in intensity of public concern rivaled that of today, occurred at that time. In the filth-riddled American cities of the 1890s, glutted by the overflow from the new industries and all the people working in them, garbage ceased to be regarded as an unpleasant nuisance and began to constitute a civic emergency.

Although collective action organized by government clashed severely with American individualism (more so than with relatively dirigist European predispositions), our first widely acknowledged pollution problem spurred a new sort of municipal activism. Crisis drove political innovation. Loosely allied in the "Progressive movement," a group of municipal reformers argued that the problem clearly could not be solved by individual action: Coping with refuse in crowded urban neighborhoods, for

---

*As in the development of industrial prowess, Great Britain preceded us in facing up to the fallout of the Industrial Revolution, as the extreme urban squalor it spawned brought the first modern environmental crisis. By mid–nineteenth century, things had gotten so bad in British cities that something had to be done, and a new phalanx of public officials and social critics went to work on the problems of the cities.

example, required organizing and paying for collective house-keeping.

When reformer Richard Ely returned to New York City from Berlin in 1879, he saw with new eyes the "dirty and ill-kept" streets, pocked pavements, and "evidences of graft and incompetence on every hand." Recoiling from the squalor that "progress" had wrought and inspired by a growing faith in the power of science and rational planning to ameliorate it, Ely and other Progressive reformers throughout the Mid-Atlantic states and the Midwest launched a concerted assault on the idea that the best government is the least government.

A confluence of crises drove home their argument. They received an ideological assist from business failures. The deep depression of 1893, during which 20 percent of industrial workers lost their jobs, seriously shook the national faith in business as social engine. At the same time, infrastructural problems affected all aspects of physical life: water supplies, sanitation, sewage, housing, and transportation. The technologies required to cope, particularly modern sewage systems, also spurred both long-term planning—for the installation of pipes, water connections, wiring, trolley tracks—and the consolidation of fragmented governmental authority into "modern" departments that could organize services. The new institutions then created for municipal maintenance made possible the flowering of urban and industrial life in the decades thereafter.

The next major surge in American governmental activism occurred in response to another massive failure in the private sector, the Great Depression. Once again, a major crisis paved the way for wholesale change in the system. After the stock market crash of 1929, massive unemployment, hunger, and homelessness forged an unprecedented acceptance among Americans of the need for government action to create the conditions for improving the general welfare. Though revolutionary in the American context, the extensive safety net of social policies erected by Franklin Delano Roosevelt's New Deal were seen by most people at the time as necessary for survival. The social spending institutionalized in that period came to seem the norm for a "modern" state.

Though running against the American grain, the value of the social safety net was not seriously contested for the next four decades. World War II and the prosperity it spurred allowed Americans largely to set aside ideological differences—at least on the home front. For several decades the public-private tug-of-war over resources was moderated by general public satisfaction. The American pie continued to swell for three decades. It was only in the 1970s that the decline in the U.S. share of world wealth began to reignite ideological passions.

As troubles multiplied, Americans fell back on traditional verities. After the Depression faded, the reasons why "social welfare" measures benefit the whole society became less vivid for most people. With growth slowing, the idea that individuals should take care of themselves and could do so much better if government didn't tax away their income gained renewed potency. Even as U.S. industry ran into serious competition from the Japanese, American public discontent focused on government intrusion, not business ineffectiveness.*

Encouraging his beleaguered countrymen to seek their future in the past, President Ronald Reagan turned his presidency into a revival of business conservatism. Failing to push back government, he implied, would mean sliding down the slippery slope toward socialism, following "those who use terms like Great Society" as a cover for greater government activism. The choice, as he saw it, was simple:

> There is only an up or down: up to man's age-old dream—the ultimate in individual freedom consistent with law and order—or down the ant heap of totalitarianism. And regardless of their humanitarian motives, those who would trade our freedom for security have embarked on this downward course.

---

*Even as average real incomes fell in 1993, reaching their lowest level since 1967, Americans did not turn their resentment on corporate leaders. While shareholder value in America's 350 largest corporations plunged by 9 percent in 1989, top executives' pay increased by 6.7 percent. During the 1980s, the pay of corporate executives climbed to 160 times that of workers. Even when people realize that the ratio of incomes of the richest 20 percent of the population to the poorest 20 percent is nine to one in the United States but only four to one in Japan, they don't seem to mind very much—perhaps because of Americans' traditional respect for private wealth.

Appealing to the preeminent value of individualism, he argued that all the money siphoned off by government had destroyed American initiative.

On a popular level the issue was joined over taxes. The fact that Americans paid less in taxes than the citizens of any other industrial nation did nothing to staunch people's indignation at giving up what they had earned.* A tax revolt that had started in 1978 with California's Proposition 13 drew substantial legitimacy from the enduring American belief that taxation is at bottom not so very different from theft of private property. (The income tax was not institutionalized in America until World War I—before that the federal government had for the most part depended on tariffs as its main source of revenue.) Reagan's clarion call "No new taxes," followed by President Bush's "Read my lips—*no new taxes!*" decisively narrowed the parameters for American political maneuver throughout the 1980s and into the 1990s.

Thus, over the past decade or so, the notion that government efforts can improve our collective quality of life has been greatly weakened. During this time, the idea grew that the key to solving intractable social problems lay in "privatization" of social services, possibly including even education and prison management. Correspondingly, Americans responded to growing social dilemmas with rising assertions of individualism. For college graduates of the 1980s the pursuit of happiness appeared overwhelmingly to mean heading to Wall Street to make your own million. A quest for security within the bounds of private property led increasing numbers of affluent suburbanites to take refuge in privately incorporated communities, walled off and guarded against intrusion.

Both the narrow equation of welfare with private wealth and the rejection of responsibility to the larger society have particularly serious implications for the environment. In the first place, the equation of welfare with wealth supports the idea that our

---

*Of the twenty-four industrial nations in the Organization for Economic Cooperation and Development (OECD), the United States is tied for last place with Greece in its tax receipts as a percentage of its annual economic output. In 1990, U.S. government receipts represented 29.9 percent of gross domestic product, as against 31.3 percent for Japan, 36.7 percent for Britain, 37.7 percent for Germany, and 43.3 percent for France.

well-being depends on economic growth. The glories of expansion are particularly potent for Americans. As we have seen, the imperative for growth is rooted first and foremost in the structure of our economy. Further, it underlies particularly American traditions of competitive boosterism, whereby every new town that sprang up on the prairie was "the first" or "the biggest" and boasted buildings that were "tallest." Images of limitless growth equate with those of endless horizons.

Second, as efforts to stem waste and pollution have intensified, the protection of private property rights has loomed as an increasingly effective bulwark against environmental protection. Gathering steam since the early 1990s, the property rights movement represents the most serious threat yet posed to the environmental reform of the last thirty years. In 1990, a group comprising miners, loggers, ranchers, and energy companies in the West and private landowners in the East and South joined forces to assert the primacy of private property rights and development objectives over protection of wetlands, waters, and endangered species or other environmental restrictions on landowners. They reasoned that one way to fight restrictions is to exact compensation for foregone development, thus making the costs to government prohibitive. In 1994, foes of environmental legislation tried to tack amendments guaranteeing compensation for restrictions on private landowners on to all proposed environmental measures.

While the grass-roots environmentalists of the 1980s found widespread public support in taking on big polluters, the property rights groups have also sought to paint themselves as a movement of the people against the powerful. In the case of the loggers versus the spotted owl, or the developers versus the wetlands, antagonists of restrictions have skillfully appropriated populist images. As they would have it, environmental curbs pit "a man and his job" against "the government" or "a man and his land" against "the power-hungry environmentalists." The U.S. Chamber of Commerce took this tack when it described the federal wetlands policy in February 1992. Casting its struggle as that of the little guy fighting for what every American holds dear, the business organization accused the EPA of operating wetlands policy "for the benefit of the environment rather than the people."

As we strive to cope with the damage that growth has wrought, we must recognize the constraints on constructive action embedded in our national mind-set. Looking at America in relation to Japan and Europe, in the 1970s and '80s the grass-roots activism that has always characterized our political life gave us a commanding early lead in confronting environmental problems. Now, however, our abiding mistrust of government makes it difficult for us to undertake the far-reaching, economywide transformation that is essential to further progress. Business as usual won't suffice—in order to stem the tide of waste we need to restructure the policies that bound the economy. And that requires government direction and action.

We will succeed in this sort of wholesale reorientation only if we play to our strengths, while also avoiding embedded ideological obstructions. Market-oriented solutions sit most comfortably within the American political context, and the market can do the job for us more effectively than any other instrument. Now we need to overcome our aversion to engineering change through public policy so that we can rewrite the rules of the game to make those solutions possible.

3

America the Bountiful

Her mighty lakes, like oceans of liquid silver; her mountains, with their bright aerial tents; her valleys, teeming with wild fertility; her tremendous cataracts, thundering in their solitudes; her boundless plains, waving with spontaneous verdure; her broad deep rivers, rolling in solemn silence to the ocean...

—WASHINGTON IRVING

There was disturbance in my heart, a voice that spoke there and said, I want, I want, I want.

—SAUL BELLOW, *Henderson the Rain King*

THE STORY OF how our waste grew starts with the saga of our material lives, of how our worldly goods have multiplied over the last two centuries. Both factory and farm production and military manufacturing have contributed an inordinate share of our garbage problem; we will take full account of their impacts in due course. First, however, we will take a look at the sweep of progress as most people see it, in the possessions that sustain their daily lives. That is mainly a story of consumption.

Our two hundred years as a nation have coincided with the greatest material change in human history. During that period, people moved from less to much, much more. Particularly in America, comfort was diffused throughout society to an unprecedented degree and consumption of this abundance increasingly shaped the way people lived.

At the time of the American Revolution, all but a few Americans lived off the land. By 1790 only 5 percent of the people lived in towns over 2,500, and only 2 percent lived in the three sizable cities of Boston, New York, and Philadelphia. Although their political revolution was behind them, Americans were about to embark on a series of material revolutions that would alter the world more swiftly than in any preceding epoch of human history. What is striking, as we look back, is how fast material life changed, after remaining relatively static for so many centuries— even millennia.

Americans seemed peculiarly suited to preside over this rapid transformation. Tocqueville's observations illuminate the readiness of these New World democrats to create a new material order. Beyond giving Americans an undoubted passion for wealth, their equality, he felt, freed their commercial spirit, for "equality begets in man the desire of judging of everything for himself; it gives him in all things a taste for the tangible and the real, a contempt for tradition and for forms." Because of the close link between politics and economics—liberty and property—in their founding ideologies, because their migration toward the frontier so strongly embodied the value of change, and because their fledgling society came closer than any other to being a virtual tabula rasa, Americans took very naturally to the material revolutions of the nineteenth and twentieth centuries.

In 1790, most Americans still lived along the East Coast. People worked at subsistence agriculture or, when there was a surplus, at processing it in mills or workshops. Although we have sharp images of our Revolution, and of its heroes riding their horses over cobblestone streets, few eighteenth-century Americans lived in a place with paved streets. Sizable halls, like that in Philadelphia where the Declaration of Independence was signed, were very few and very rude by European standards. More crucially, the tools and furnishings people used in their daily lives, whose quaintness we relish, were also few. What we cannot see is what Americans *didn't* have in 1790.

Thomas Jefferson was fascinated with machinery. His home at Monticello boasted ingenious pulleys, hinges, winches, turntables and pumps that caused panels to recede, doors to open without being touched, writing desks and closets to revolve and food to arrive in the dining room without servants. Yet he saw his gadgets chiefly as entertaining toys—displays of human ingenuity—and certainly did not view them as potentially marketable products. Cultural historian Leo Marx has pointed out that Jefferson in fact visualized commercially operative machinery as situated in the midst of bucolic rural landscapes—a gristmill beside a sparkling stream providing a favorite image. Machines in great factories manufacturing quantities of goods for domestic consumption were beyond most people's imagining.

For us, on the other hand, it requires a real effort to picture how few possessions our forebears had to serve their everyday needs. In eighteenth-century America, where transportation and trade networks were still forming and manufacturing had been limited by the Crown, people were more self-sufficient than in Europe. Tools were essential to make or grow what one needed, but most people had few. A farmer needed hoes, axes, sickles, a plow, which might be borrowed, a gun, for obtaining fresh meat. Some might also have harnesses for livestock, pails for milking cows, buckets and spigots for tapping maple trees, and wagons or buggies for transport. Seeds, manure, and compost were culled from growing cycles and residues. Some labor-saving devices and conveniences could be purchased from metalworks, village artisans, or peddlers, but the meager cash income of the ordinary farmer made metal implements and other manufactured goods prohibitively expensive, compared, for example, with land or building materials.

Within the farmhouse or rural cottage, furnishings varied according to wealth and locale, but in general, most items of daily use were manufactured by the farm family. When not cooking, cleaning, or doing the laundry, wives and daughters produced what the household consumed: by churning butter and setting cheese, brewing beer and baking; rendering fat for soap and molding tallow into candles; and sewing and knitting clothing as well as spinning or weaving if they owned a loom. Male household members also produced a number of tools in addition to splitting and piling fuelwood, taking grain to the gristmill, butchering animals for meat, and fashioning furniture and household implements such as brooms, brushes, wooden trays, bowls, platters, lamps, buckets, dye tubs, churns, doormats, baskets, bread troughs, and washboards.

Diets were not highly varied: Ordinary people ate bread and porridge; eggs, butter, cheese, and milk when they could; vegetables and fruits in the summer and preserves in the winter; and meat when it was available. Social historian Ruth Schwartz Cowan evokes a telling image of eighteenth-century domestic life, and the tools that usually sustained it, centering on the ubiquitous stew that served as the main meal for most people most of the time.

Cowan cites a recipe from Albany in the 1780s, the principal ingredient of which was a hacked up leg of mutton, set over the fire in a pot of water. When the pot came to a boil the cook would add chopped "violet leaves, endive, succory, strawberie leaves, spinage, langdebeefe, marygold flowers, scallions and a little parsley" mixed with chopped oatmeal, boiled until "the meat be boild enough," and all well mixed together, to be served with a pinch of salt. To concoct this dish, the husband would have butchered the sheep with knives made of wood and iron; husband or wife would have carried the water from stream or well in a wooden bucket (perhaps bound with leather or iron); the stew would cook in an iron kettle suspended over a fire resting on brass or iron andirons.

Of the ingredients only the salt was purchased; everything else was grown on the farm. During the cooking process, the housewife stirred the stew with wooden spoons whittled by her husband, and served it in wooden trenchers carved by him. When the family finished eating she would wipe the dishes with a rag and clean the kettle with some water and sand.

While supplying the simplest necessities often required laborious effort, maintaining them was much less complicated and time-consuming. There were no bathrooms, stoves, or refrigerators to scour. Before the advent of affordable cotton cloth, laundry was also a sometime thing. In sparsely furnished rooms with bare wooden floors, it would usually suffice for the wife to sweep the planks with a broom she had made, or scrub them with lye of her husband's manufacture. Garbage was the least of her problems; scraps were fed to the animals and the family ate any leftover bread or stew shortly thereafter, probably at the next meal.

Many Americans, especially those who left East Coast amenities to strike out for the forests and prairies of the frontier, lived similarly close to simple subsistence far into the nineteenth (and even the twentieth) century. Power was supplied by wind, water, animals, or humans. After the Revolution, craftsmen in the cities and towns produced metal goods, furniture, textiles, and shoes for those who could afford to buy them. Mills turning out ground grain, lumber, and ironware employed machinery at a larger scale. But even in the mills, all the machines were handmade. Except for the shaft bearings and spike or clamp fasteners, the machinery in the water-powered mills, even the gears, was made of wood.

Because waterpower was dispersed throughout the countryside and artisans worked in their homes or workshops, cities had not yet acquired their later importance as centers of production. Only two American cities had populations exceeding 25,000 in 1790, and decades later, even the largest cities retained some of their rural character: As late as 1850 in New York City, boys could still rope pigs in the streets, and fish in ponds scattered about Kip's Bay.

Though American metropolitan areas at that time were less squalid than Europe's older urban centers, and particularly Britain's burgeoning factory cities, their sanitation technologies were very basic, differing little, in fact, from those of rural areas. People drew their water from wells and used privies in backyards and alleys. Collection systems were few and rudimentary, so refuse and slops often ended in the streets, where the principal means of disposal were scavengers and rooting pigs. For the most part, however, sanitation and garbage in preindustrial cities could be seen as problems for individuals, to be dealt with according to wealth, poverty, neighborhood.

Historian Harold Burstyn aptly evoked the enormous material gap between the Founding Fathers' America and that of the mid–twentieth century with a story about George Washington's view of the site that would later be the city of Pittsburgh. On a horseback trip through the back country of Virginia and Pennsylvania in 1753, he came to the place where the Allegheny River, running south, joins the Monongahela, flowing north, to form the westward-bound Ohio.

On that low, marshy, triangular piece of ground, where now stand the handsomest and tallest buildings of the city of Pittsburgh, Washington found only trees. Upstream along the Allegheny, where now a modern industrial city has its factories, hospitals, warehouses, and sewage plants in an unbroken row, Washington could see only the clearing that marked a trading post. The banks of the Monongahela, marked now by blast furnaces and rolling mills on the right bank, open-hearth steel furnaces on the left, was similarly empty of Western civilization except for another trading post.

Obviously Washington could not have envisioned the Pittsburgh of two centuries later, where he saw two trading posts in the forest. Yet by the end of his lifetime, and of the eighteenth century, the nation was entering an epoch of extraordinary technological change. Already a generation of inventors and gifted mechanics was laying the groundwork for the factory production and steam-powered transportation that would shortly bring into view the world as we know it.

Until around the mid-nineteenth century, the majority of American mills and factories still depended on rushing streams, rivers, or waterfalls for power; thus they were dispersed in relatively isolated rural settings, like most of the nation's population. European travelers like Harriet Martineau marveled at the bucolic setting in America for factory production, which in Europe consigned workers to squalid city life. On a trip to this country in 1837, she was struck by the good fortune of American workers, with "their dwellings and their occupation fixed in spots where the hills are heaped together, and the waters leap and whirl among rocks." Her pastoral factoryscape aptly represented the predominantly rural and agrarian society that was then America.

From the 1850s to the 1880s, the nation's nascent industrial establishment made the transition from waterpower to steam power. Overlapping the transition in technologies from water to steam was a shift in fuels from wood to coal. The use of wood as fuel for steam power hastened the decimation of wood supplies and helped push up costs in depleted areas. Correspondingly, the move from water- to steam power made possible the concentration of industry in the cities and drew more people there, increasing the pressure on wood supplies in populated areas.

Until the late nineteenth century, wood was the foundation of the American economy, not only for fuel, but as the chief construction material—for houses, furniture, ships, carriages, wagons, and tools—as well as the source of chemicals like potash and tannin, and of charcoal for ironmaking. Exploitation of forest resources, however, eventually brought shortages and escalating costs to Eastern cities.* As early as 1841, the scarcities gave rise to a conservationist anxiety about forests that sounds much like today's environmental concern:

> In the old states men are beginning to estimate trees rather as timber than fuel, and the time is rapidly passing away, in all parts of the Union, when it is deemed that the best mode of disposing of the noble trees that grace the American woods is to turn them into ashes. On the sea-coast, that time has long since past, and for many years the community has been anxiously seeking some substitute for the rapidly diminishing forest.

As usual in American history, the Lord provided a new option—in abundance. Coal, called by *Scientific American* in 1853 "our black diamonds," became a ready substitute for wood in heating cities and fueling industrial expansion. But where the production and distribution of cordwood had been, as historian Louis C. Hunter has remarked, "quite literally a matter of individual enterprise"—a job for a farmer or woodsman with an ax and a wagon—mining, with its underground construction and aboveground marketing and distribution, meant heavy capital costs and industrial organization on a new scale.

The synergies between coal mining and the railroads, and between the railroads and industry, changed the nation very dramatically in the decades of the mid-century. From its inception, rail transport provided the key link between mines, ironworks, cities, and industries in different regions, making possible manufacturing and commerce on a new scale. Expanding from 73 miles of track in 1830 to 30,636 in 1860, the railroad altered land travel irrevocably, thereby profoundly altering the land itself. Within those thirty years, railroads threaded the continent together. At the same time, they laid the groundwork for a huge

---

*Because of our bounteous endowment of woodlands, Americans turned from wood to fossil fuels considerably later than the Europeans, whose forests had been denuded by centuries of exploitation. In the 1850s the average American family still burned eighteen cords of wood every year (about equivalent to 2.5 tons of coal per person).

national market that was fostered and fed by the new industries that also battened on steam power.

For Americans, the powerful presence of the railroad, shooting and snorting across the land, kindled a passionate love affair with the machine. Its movement and speed etched across the national landscape was a manifest fulfillment of shared hopes for material progress. By 1840, when 1,818 miles of track had been laid in Great Britain, there were already 2,800 miles of it in the United States. A French visitor saw the locomotive as the personification of the American: "The one seems to hear and understand the other—to have been made for each other—to be indispensable to the other." Americans themselves wrote incessantly about the glories of the railroad, as in a flowery passage of 1853 in *Putnam's* magazine:

> And the Iron Horse, the earth-shaker, the fire-breather, which tramples down the hills, which outruns the laggard winds, which leaps over the rivers, which grinds the rocks to powder and breaks down the gates of the mountains, he too shall build an empire and an epic. Shall not solitudes and waste places cry for gladness at his coming?

From mid-century, railroads, factories, and cities all grew in concert. The interrelationships are obvious: With the factories, workers came to the city; with the railroads, the city came to the hinterland. Moving crops to market and bringing supplies to the farmers, the railroads quickened the economies of old commercial centers such as New York, Philadelphia, Cincinnati, Cleveland, and St. Louis and spurred the creation of new cities like Chicago, Indianapolis, Atlanta, Kansas City, Dallas, and Minneapolis–St. Paul. By 1880 nearly 30 percent of Americans lived in cities, and these growing urban masses were consuming manufactured clothing and processed food in increasing quantities.

Arguably the most far-reaching technological revolution, though, was through electricity. Electricity began to change people's lives toward the end of this period, even though its widespread transformation into heat, illumination, and power lay ahead in the twentieth century. The telegraph, which involved a relatively simple basic technology and form of transmission, came first. By the late 1840s more than two thousand miles of telegraph lines had been strung, in tandem with the railroad. By the end of the Civil War telegraphic links had become essential for ensuring safe passage, on schedule, via rail. Without realizing it, people were becoming prisoners of precision.

As Americans emerged from the social dislocation of the Civil War, the Industrial Revolution was beginning to change their world. Over the next several decades, long-established patterns of economic interchange altered to meet the new conditions. Whereas international traders had previously dealt principally in raw materials, now they purveyed manufactured goods. Bankers revised their forms and standards to accommodate capitalists who wished to invest not in land but in machinery. Storekeepers learned to exact payment in cash rather than in kind. And the enterprising began to think about patenting a bright idea or seeking technical training rather than acquiring a homestead.

Everyday life changed most dramatically in the cities. Population growth soared in the United States. Between 1850 and 1920, American cities grew up to six or seven times faster than those in the rest of the world. After 1890, American farmers streamed to factory jobs in the city, accompanied during this period by massive waves of immigration, mostly from southern and eastern Europe. Between 1840 and 1850, the numbers of people in cities increased 92.1 percent, and grew again tenfold between 1850 and 1890. In 1894 part of New York City passed Bombay, India, as the most densely populated area of the world.

This population explosion severely challenged the habitability of urban streets and structures largely devoid of sanitation and public services. Charles Dickens wrote of a New York slum in 1850:

> See how the rotten beams are tumbling down, and how the patched and broken windows seem to scowl dimly, like eyes that have been hurt in drunken frays. . . . Where dogs would howl to lie, women and men and boys slink off to sleep, forcing the dislodged rats to move away in quest of better lodging.

The rise in infant mortality rates in New York City, from 120 per 1,000 in 1810 to 240 in 1870, gives stark evidence of declining sanitation and health conditions. Some decades after the factory had, according to Lewis Mumford, transformed British industrial towns like Manchester "into dark hives, busily puffing, clanking, screeching, smoking for twelve and fourteen hours a day, sometimes going around the clock," American cities suffered industry's blight. A century after Washington's visit to the woodland site that became Pittsburgh, that mining and smelting center was, according to an English visitor, so thoroughly coated with coal dust that it left "every fresh green leaf and the delicate flower . . . begrimed ere they have fully unfolded themselves, by the smoke and soot with which the whole atmosphere is impregnated."

Though Pittsburgh in its steel-producing heyday represented an extreme, the heady growth of manufacturing nationally brought new sorts of dirt, trash, and toxic waste to the cities where the factories clustered. In the last half of the nineteenth century, the overall volume of production increased thirtyfold, and between 1859 and 1899 the output of industrial plants rose from $2 billion per year to more than $11 billion. The census of 1880 listed a number of industries that had been brought under the factory system during the previous thirty years, including "the manufacture of boots and shoes, of watches, musical instruments, clothing, agricultural implements, metallic goods generally, fire-arms, carriages and wagons, wooden goods, rubber goods, and even . . . the slaughtering of hogs." Between 1867 and 1897 the yearly production of iron and steel increased nearly five times, from 1.6 million tons to 7.2 million tons.

Not surprisingly, metropolitan governments were unprepared to deal with the resulting pollution and debris. Heretofore, smoke had always been the sign of a prosperous community, and the prerogatives of factories had for the most part superseded the needs of human inhabitants. As a rule, the new factories laid claim to choice sites along the waterfront and exercised a nearly total lien on the urban environment. In *The City in History*, Mumford summed up the consequences:

> For generations, the members of every "progressive" urban community were forced to pay for the sordid convenience of the manufacturer, who often, it happened, consigned his precious by-products to the river, for lack of scientific knowledge or the

empirical skill to use them. If the river was a liquid dump, great mounds of ashes, slag, rubbish, rusty iron, and even garbage blocked the horizon with their vision of misplaced and unusable matter.

Motivated by a widespread perception of crisis and the zeal of the Progressive reformers, the cities embarked on massive investments in infrastructure (much of which is still in use today, though under considerable stress). Planning and building continued for decades, weaving pipes, tracks, and wires over and under avenues and alleys. In the 1890s, after railroad construction had been essentially completed, the cities became the major U.S. market for the steel, copper, and heavy machinery industries.*

Riding the wave of industrial expansion, Americans—at least in the upper strata—grew wealthier. At the end of the nineteenth century, American firms, in great part still family affairs, were combining, consolidating, merging, and reorganizing, paving the way for the advent of the giant corporation. Yet while manufacturing wealth was increasing, between 1870 and 1900 the gap between rich and poor widened. Although the American standard of living and growth in per capita income had been among the highest in the world since the end of the eighteenth century, a typical family in 1875 still used more than half its income on food and almost all of the rest for a subsistence level of clothing and shelter. Only in the middle and upper class, particularly in urban and suburban areas, had material life changed.

Within the homes of the affluent, life acquired a new gloss. Home was viewed as a kind of sanctuary, and domestic arts and accessories became the object of almost obsessive attention. Nu-

---

*Industrial historian Alfred Chandler has pointed out in his pathbreaking studies of industrial strategy and structure that after this time municipal infrastructure ate up growing quantities of electric lighting, telephones, copper wire, newsprint, streetcars, coal, metals, piping, structures, and fixtures. Andrew Carnegie signaled the change in the market for producers' goods in 1887, when he shifted the output of the Homestead Works, then the largest and most modern American steel plant, from rails to structures.

merous books and articles, including among their authors intellectual luminaries such as Harriet Beecher Stowe and her sister Catherine, lavished attention on household cleanliness, fine "cookery," and interior decoration.*

As urban historian Joel Tarr has pointed out, during this period prior to the final years of the century, most urban dwellers still drew their water from ponds and streams, wells and rainwater cisterns. They disposed of wastewater from cleaning or cooking by throwing it on the ground or into the gutter, a dry well, or a cesspool. Human wastes were deposited in cesspools or privy vaults close beside residences, which allowed wastes to seep into the soil (and into wells and groundwater) or, if they were impermeable, required frequent emptying.

From mid-century onward, but chiefly during the 1880s and 1890s, municipalities worked to replace often polluted wells and cisterns with pure water piped in from rivers, streams, and res-

---

*Among domestic virtues, "purity" ranked high, and was reinforced by the "filth theories," about the relationship of dirt and disease, currently shaping thinking about public health.

Cooking was probably held in higher esteem in that period than in any other era of American life except, perhaps, the heyday of gourmet cuisine during the 1960s and '70s. The literature of the period spends considerable time on domestic artistry. In Uncle Tom's Cabin, for example, Stowe contrasts the bright sunshine of domestic happiness with the darkness of slavery, when, in one characteristic passage, Eliza watches the Indiana Quaker family absorbed in making breakfast:

> For a breakfast in the luxurious valleys of Indiana is a thing complicated and multiform, and, like picking up the rose-leaves and trimming the bushes in Paradise, asking other hands than those of the original mother. While, therefore, John ran to the spring for fresh water, and Simeon the second sifted meal for corncakes, and Mary ground coffee, Rachel moved gently and quietly about, making biscuits, cutting up chicken, and diffusing a sort of sunny radiance over the whole proceeding generally.

Notable gurus rhapsodized in print about, for example, tiles in "symphonies of red and blue," vines and flowers growing around picture frames, or the disadvantages of white ceilings and white woodwork (which could "put a room out of tone"). As a result, male as well as female members of the family devoted serious consideration and scrutiny to domestic furnishings and equipment. The optimum arrangement of kitchens was much discussed and pictorially represented in household manuals. And as cherished accessories of family life, household goods generally acquired value transcending the utilitarian.

ervoirs. Although Philadelphia built a waterworks as early as 1802, New York was the first big city to achieve an ample supply, bringing clear water from upstate New York to the Croton Reservoir and thence to the city via aqueducts completed in 1842. By 1860 the nation's sixteen largest cities had waterworks, and by 1880 the number of waterworks had increased more than fourfold. Piped water now flowed into middle-class homes.*

Although electrification first appeared widely in arc lighting for city streets in the 1880s (which replaced gas lighting installed in the 1820s, '30s, and '40s), household electrification began slowly. As late as the Philadelphia Centennial Exhibition in 1876, the public viewed electricity as important principally for telegraphy and electroplating. The full-scale adoption of electricity came more than twenty years later, with the extension of cables through cities and their outskirts, home wiring and metering, and the replacement of kerosene and gas fixtures with new electric lamps designed around sockets for the light bulbs. Equally, people did not yet grasp the potential of the telephone, seeing it as an application of telegraphy useful for the transmission of public speeches. In the words of the July 1876 *Boston Transcript,* "If the human voice can now be sent over wire . . . we may soon have distinguished men delivering speeches in Washington, New York or London, and audiences assembled in Music Hall or Faneuil Hall to listen." Rates were high—in 1895 telephone service still cost $4.66 a month, equivalent to 13 percent of an average worker's wages—and it was impossible for most people to imagine using the telephone just to talk. Once rates decreased, however, the telephone became for Americans a natural extension of themselves.

From 1900, when there were about two telephones for every one hundred people, usage grew to around sixteen per one hundred (and 42 percent of households) in 1929. Like the railroad and the telegraph, the telephone offered a new sense of control in the conquest of time and distance. It also catered to the endemic national impatience, what Tocqueville had described as

---

*It really had no adequate place to flow *out,* however, until sometime later, as centralized water supplies preceded sewerage systems by a decade or more— and greatly increased the need for them, when people began to connect their new water closets to existing storm-water runoff gutters.

"permanent agitation," by speeding things up in several ways—allowing people to bypass the tedious process of writing down their thoughts and enabling them to transmit their impulses immediately. The age of electronic communication was beginning.*

All of these "modern inventions" flourished first in the nineteenth-century hotel. Historian Daniel Boorstin has catalogued the many mechanical comforts and gadgets tried first in the palatial commercial lodgings and gathering spots peculiar to this country at the time. Boston's Tremont House, built in 1829, pioneered in extensive plumbing with a battery of eight water closets on its ground floor. A few years later, New York's Astor House astonished guests with plumbing on the upper floors, fed from a roof tank to which water was pumped by steam power. The first American building partly heated by steam was Boston's Eastern Exchange Hotel (1846). In 1882 (shortly after Edison produced his light bulb), New York's Hotel Everett became the first hostelry lit by electricity, with 101 incandescent lamps in its main dining room, lobby, reading room, and parlors. The telephone became a common hotel fixture after New York's Hotel Netherland installed room phones in 1894.

Another new form of urban habitation, the apartment building, emerged as a direct result of all these material innovations. By the late 1880s rising land values and technological possibility combined to push residential buildings upward (as they had office buildings a bit earlier). Historian Glenna Matthews has shown how the apartment offered unparalleled convenience to its middle-class inhabitants, with piped in water, refrigeration, and gas, as well as elevators "to move the tenant and her groceries." The latest in kitchen design undoubtedly had particular appeal to housewives who had grown up with pumped water and wood stoves:

> The floors were typically covered with rubber tile, and the lower half of the walls had white tile that was easy to clean. A gas stove,

---

*In our own time, the computer and fax, which allow the written word to be transmitted instantaneously, may stimulate something of a revival for written personal communication. For the reticent, a brief written message may come more easily than one blurted into a telephone recorder. Whether computers and faxes will bring a return to the lengthy epistolary exchanges of the past, however, is more doubtful.

a hot water heater, a sink with hot and cold running water, and a refrigerator with "glass or tile linings and compartments for every conceivable use" were all in place, simplifying almost every household task.

By 1880 about a quarter of urban dwellings boasted a bathroom (commonly known as a "water closet"), and by the end of the century this facility was becoming standard for middle-class apartments, though not for tenements.* The full bathroom would become commonplace only after the First World War.

By the end of the century, the domestic lives of the urban and suburban middle classes diverged widely, not only from the lives of the poor, but also from those of people in villages and on farms, whose homes did not yet have the same access to piped water, sewerage, gas, or electricity. Even skilled workers by and large accumulated few possessions and had little money to spend on leisure-time pursuits. Nonetheless, some changes affected material life more widely. Cooking moved from hearths to cast-iron wood stoves in a great many homes after 1830, and innovations in the iron and steel industry helped bring the price of a stove within the reach of ordinary people by mid-century.†

With improved home preserving equipment women continued to "put up" quantities of fruits and vegetables, but during the eighties and nineties advances in packaging, particularly commercial canning, brought Campbell's soups, Heinz pickles, and a variety of prepared foods, tinned milk, and packaged butter. Most households ceased to butcher their own meat and, when they could afford it, bought the products of the meatpackers in Chicago. By mid-century even farm families had switched from grinding home-grown wheat, rye, and cornmeal to buying superfine milled wheat flours. Thereafter, growing numbers bought

---

*Although domestic guru Catherine Beecher's 1869 design for a city flat included a bathtub, piped water came last to the bath—after the kitchen sink, the washbasin, and the toilet.

†At mid-century stoves ranged in price from five dollars to twenty-five dollars, while laborers made about a dollar a day.

their bread at a bakery, and by 1900, 90 percent of urban homes relied solely on bakery bread.

Although most clothing was still sewn at home, women had been fashioning their families' wardrobes from manufactured cloth since the 1830s, aided by sewing machines since the 1860s. As a result, people's wardrobes expanded. Into the next century, the task of doing the laundry—whose mainstays were the tub and washboard—would become increasingly onerous, as people acquired more clothes and the advent of cheap cottons greatly increased their washability.

Household objects were not yet purchased cheaply, and consequently they were often cherished through their owners' lifetimes. Gradually, though, they were becoming available in increasing variety. Even before electrification, the array of new household conveniences, ranging from eggbeaters and carpet sweepers to porcelainized cookware, would have fascinated and amazed Thomas Jefferson. As early as 1850, for example, a Philadelphia hardware store offered 250 sorts of kitchen tools.

To market this profusion of goods, new forms of merchandising were needed. Their common denominator was the goal of high-volume sales deriving from the availability of large quantities of goods purchased for cash. The volume of goods when combined with increasing self-service often led to lower prices. The general store, itself an American invention and a fixture since the Revolution, continued to hold sway in rural communities, even as its descendants, the supermarket and department store, were taking hold throughout the East and Midwest.

An early precursor of the supermarket, the A&P (originally, the Atlantic & Pacific Tea Company) was founded in 1859. Although operating still on a relatively small scale, the new emporium aimed to provide lower prices for high-volume sales of every foodstuff housewives might need or fancy, including groceries, meats, vegetables, and fruits. The beginnings of the department store are more ambiguous: Alexander Turney Stewart, a native of Belfast, is believed to have opened the first department store in this country, known as the Cast Iron Palace, at Broadway and Ninth Street in New York City, in 1862. Macy's was also an early prototype, evolving gradually as R. H. Macy added one specialty shop to another and then another until he had several "departments" under one roof, around 1860.

In 1872 Aaron Montgomery Ward used the same merchandising principles to tap a more dispersed market when he published a one-page circular offering a variety of mail order goods from a shipping room in Chicago. Serving a far-flung clientele, the mail order businesses succeeded the ubiquitous itinerant peddlars who had wandered rural lanes since colonial times. Before twenty years had passed virtually everything, from bonnets to bannisters to buggies, could be ordered from the catalogs. In addition, as the century wore on and national distribution networks powered by the railroad took more definite shape, a new breed, the traveling salesman, arrived en masse to service them. Numbering more than six hundred thousand by 1911, these men-on-the-go, or "drummers," became emblematic of the coming age of "salesmanship."

On the other side of the counter, the advent of the department store heralded a new popular pursuit, "shopping." Particularly for urban and suburban women of the "leisure class," with husbands at work, children in school, and servants in charge of the household, the department store became entertainment and avocation. Even more, it was the temple of the new prosperity. Stewart's store, eventually occupying eight two-and-a-half-acre floors, was staffed by more than two thousand employees and featured continuous organ music to attract and entertain customers. Historian Genna Matthews comments on the "cathedral-like" character of New York's Macy's, Philadelphia's Wanamaker's, and Chicago's Marshall Field's, noting that none other than President William Howard Taft extolled them as a source of incomparable human value, for bringing together

> ...under one roof...at the lowest reasonable, constant, and fixed price, everything that is usually needed upon the person or in the household for the sustaining of life, for recreation, and for intellectual enjoyment.

They were designed as pleasure palaces and social centers. Wanamaker's, for example, contained an auditorium seating two thousand, an organ with nearly three thousand pipes, a marble-colonnaded Grand Court, and a Greek Hall with six hundred more seats, while at New York's Siegal & Cooper Company the entrance featured a fountain playing over a twenty-foot-high goddess. Adjoining Macy's Louis XV–style ladies' waiting room was an art gallery exhibiting a collection of oil paintings. In addition to its clublike reading and letter-writing facilities, Macy's provided special promotional activities like bicycling lessons. Many stores also opened ladies' lunchrooms. Thus, as Marion Talbot, dean of women at the University of Chicago, observed in 1910, they offered a major social outlet, taking the place in women's lives of the "sewing circle, the husking bee, the afternoon visit."

As we have seen, the most fundamental shift in middle-class lifestyles, which laid the groundwork for sweeping changes in national life during the next century, occurred around the same time, with the advent of the suburb. In the decade after the Civil War, Americans who could afford to began to move from the city to the new planned communities like Evanston, outside Chicago, and Llewelyn Park, across the Hudson River from New York City. As the suburbs became the main beachhead for upwardly mobile professionals of the day, their growth spurred a wave of consumption beyond the towns' expenditures on infrastructure. For families, building, equipping, and decorating their separate enclaves became a joint activity and major preoccupation. Merchandisers catered to the convenience of this favored upscale market. Suburban housewives took the train to the city to shop in the department stores and then returned home to await delivery of their purchases— often on the same day. Husbands could have packages that they had purchased in the city during the day delivered to the railroad station to be picked up at the parcel room on their way home.

The new way of life created a demand for all sorts of new equipment. Self-sufficiency dictated that each family purchase a complete set of everything. With electrification came more household appliances aimed at supporting higher and higher levels of domestic proficiency.* Eventually, escalating standards of cleanliness sparked whole new industries. Doing it yourself, always an American ideal even before it became an advertising slogan in the 1950s, also required the latest in tools and gadgets. Moreover, every home needed to have its own landscape. Even then, the American preoccupation with the lawn was transmuting gardening into a high order of civic responsibility, with concern for one's crabgrass measured as consideration for one's neighbors. Agricultural technologies and tools aimed at evening and weekend horticulturalists in Scarsdale, Shaker Heights, or Winnetka started with horse manure and the rake but grew progressively more elaborate. Self-sufficient suburban life also required mobility. As the new century got underway, that came to mean, above all else, "a car in every garage."

When the United States passed Great Britain around 1900 as the world's leading industrial power, the early-twentieth-century captains of industry were inventing new forms of business organization appropriate to the national scale of their corporations. Paralleling the technological breakthroughs occurring in manufacturing and chemical engineering, they effected an organizational revolution that made the production boom possible.

Meanwhile, on the factory floor, the most famous single example of "scientific management," Ford's assembly line, revolutionized the industrial process. As we have seen, Ford perfected volume production. Where in 1899 existing automobile manufacturers were struggling to produce a few cars a week in barns, repair shops, or improvised plants, by 1925 Ford's process was turning out a car every ten seconds. Dropping in price from $850 in 1908 to $290 in 1925, the Model T or "flivver" generated enormous profits for the automobile and related industries, spurred the development of petroleum extraction and refining over coal mining and the road system over mass transit, and changed national life by making private transport affordable for ordinary people.

Even more than the railroad, the automobile brought a heady sense of technological progress and extended sway over nature

---

*As more conveniences were becoming available to ease onerous tasks, however, servants were becoming scarcer. Middle-class housewives found themselves directly responsible for more of the mundane cooking, cleaning, and laundry.

because it freed people to transport themselves. Its effects on American life have been manifold and well documented. As a cultural icon it has probably had no equal in our two-hundred-year history. Sinclair Lewis summed it up in 1922 when he wrote of the hero of his novel *Babbitt*, and of the other well-off citizens of George Babbitt's city, Zenith, that "his motor car was poetry and tragedy, love and heroism." In the nation's material evolution—its industrial growth and the transformation of the domestic marketplace, the spread of the suburbs, the use of leisure, the generation of waste—the car's impact has been unparalleled.*

The growth of the American auto industry astonished the world. Production of cars and trucks climbed to 5.6 million in 1929, nearly five sixths of world output, making possible 1 car for every 5 people in the United States (compared with 1 for 43 in Great Britain, 1 for 325 in Italy, and 1 for 7,000 in Russia). Given the automobile's cost, about 20 percent to 45 percent of average incomes, the speed with which Americans embraced it was breathtaking. By 1930, for thirty million households nationwide there were twenty-six million registered automobiles. Since World War II the industry has been the country's largest single source of employment and gross domestic product (GDP). Even in 1929 one tenth of all nonagricultural labor worked at making automobiles.

According to historian Martin Melosi, auto manufacturers bought 20 percent of the steel, 80 percent of the rubber, 75 percent of the plate glass, and 25 percent of the machine tools America manufactured in 1930. The auto industry also spawned dozens of ancillary businesses, like parts manufacturers, tire dealers, and service stations. Closely aligned endeavors included land development and highway construction. Like the railroad, the auto expanded the access to the land that Americans had always considered their birthright. Newly paved roads smoothed the way to suburbs, farms, the countryside, and travel across the nation.

For people generally, nearly universal access to private trans-

---

*The gasoline-powered internal combustion engine was developed earlier than the auto. The first was built in Germany in 1876, and by the 1890s the basic design of the automobile engine still in use today had been perfected. Steam- and electric-powered engines competed briefly with the internal combustion engine, but after the massive 1901 oil strike at Spindletop in south Texas (then the world's largest producing well outside the Baku field in Russia), the synergy between the automobile and petroleum industries became a fundamental fact of twentieth-century American life.

port represented a striking social rearrangement. In the era of horse-drawn vehicles only the very rich could afford private carriages: In their initial study of Muncie, Indiana, *Middletown*, the sociologists Robert and Helen Lynd reported that in the 1890s, only 125 families owned a horse and buggy, but by 1923 there were 6,222 passenger cars, or two for every three families in the city.

Moreover, Americans' fidelity to their automobiles was intense. During the Depression, people who gave up almost everything else clung to their cars. The Model T was so simply constructed that an average person could often repair it with twine, wire, clothespins, or whatever else came to hand, and half its parts could be purchased at a garage for less than fifty cents. In their second, post-Depression, *Middletown* study, the Lynds pointed out that while retail dollar volumes in Muncie fell 30 percent to 70 percent between 1929 and 1933, filling station sales fell only 4 percent. Moreover, car registration per capita in that city between 1932 and 1935 remained above the 1929 level. As they observed, car ownership was "one of the most depression-proof elements of the city life in the years following 1929—far less vulnerable, apparently, than marriages, divorces, new babies, clothing, jewelry."

The story of the automobile sums up the material history of America during most of the twentieth century. Both industrial production and people's lives were shaped by its progressive *diffusion* throughout society. Escalating productivity, accompanied by rising wages and falling prices of manufactures, eventually enabled most Americans to enjoy the comforts that had formerly characterized a "middle class" standard of living.

In a real sense, Americans in the nineteenth and early twentieth centuries laid the technological and institutional groundwork for the material harvest that their countrymen would reap in the second half of this century. World War I offers a convenient demarcation point to measure the changes. For Europe's privileged classes the war sounded the death knell for an old social order; for Americans, its end signaled the beginning of a new economic era, based on the imperative for consumption.

In 1914, for the first time more than half of the U.S. population lived in urban or suburban areas. By 1919 the country was entering a period of unprecedented prosperity, which would last until the crash of 1929. The gross national product bounded upward from about $73 billion in 1920 to about $103 billion in 1929. With corporate growth, a burgeoning group of middle-level managers expanded the ranks of the middle class. In the 1920s wage rates rose three times as fast as they had in the decades since the Civil War, and industrial workers' wages rose 11 percent to 13 percent. (In Muncie, the Lynds reported, incomes grew twice as fast as they had before the war.)

To be sure, the degree to which the rising tide lifted all boats should not be exaggerated: Less than half the population was "comfortably" off, or able to enjoy most of the new amenities that characterized the era. In 1920 few workers made more than $600 a year, and by 1926 average buying power was about half that of today. Nonetheless, the massive wartime inflation narrowed income gaps somewhat by raising the relative wages of unskilled workers. Moreover, prices of electricity and appliances had continued to drop since 1900 to levels much lower than those abroad. Thus the quantum variations in levels of comfort available to different income levels were beginning to blur.

At the same time more of the work force gained more leisure time. The work week had dropped from sixty-five hours in 1850 to sixty in 1890, to fifty-five in 1914. In 1923, U.S. Steel moved from the twelve-hour day to an eight-hour shift at its Gary, Indiana, plant, and in 1926 Ford instituted a five-day work week. By 1930, manufacturing workers averaged a 42.1-hour week. Middle-class families began to explore the custom of "the vacation," which had been virtually unknown to the previous generation. The automobile, of course, stood ready to transport its owners to the new world of the pleasure trip. The atmosphere of American material life was changing. Although it would receive a brutal setback with the 1929 crash, the consumer revolution was underway.

The view from the top down during the decade showed a major displacement of corporate focus: Whereas in 1919 only one of the leading twenty U.S. corporations manufactured consumer goods, by 1930 nine of the twenty leaders aimed at the consumer market. On the demand side, a number of important things changed. Consumer credit expanded enormously, and so did consumer debt. By the mid-1920s, over 75 percent of new autos, 70 percent of furniture, 75 percent of radios, 90 percent of pianos, 80 percent of phonographs, and approximately 80 percent of all household appliances were purchased on credit. During the decade consumer debt doubled. Conversely, consumer savings rates fell by one third.

The burgeoning availability of credit made possible the enormous expansion in consumption. Economist Martha Olney argues convincingly that the expansion of consumer credit was not merely a simple market response to pricing and demand but rather a deliberate marketing strategy of automobile and appliance manufacturers. In the 1920s a large proportion of both automobile and electrical appliance sales were made on installment plans arranged by finance companies that were usually tied very closely to the manufacturers. Given the size of these purchases in relation to incomes, the financial innovation of installment buying was essential to expand markets and smooth distribution.

General Motors' Alfred P. Sloane, Jr., whose marketing innovations we noted in Chapter 2, set up his own finance company in 1919. Characteristically, Henry Ford resisted this trend, trying unsuccessfully to get consumers to lay away their car's purchase price in advance. Finally, however, in 1928, he gave way and followed suit. By the end of the 1920s General Motors had overtaken Ford to assume the lead among U.S. auto manufacturers it has held ever since.

In 1845, the U.S. patent commissioner, Henry Ellsworth, thought the Patent Office should be closed because everything useful had already been invented. He would have been flabbergasted had he been faced with the profusion of consumer goods, many based on corporate scientific research, that arrived in American stores in the 1920s: lamps, record-players known as "victrolas," radios, washing machines, vacuum cleaners, and refrigerators from the electrical industry; materials such as rayon and cellophane, Pyrex and plastic utensils, linoleum floor coverings, antifreeze and cosmetics from the chemical industry. All in addition to the affordable automobile! As the historian Siegfried Giedion put it, from 1918 to 1939 "full mechanization penetrated the intimate spheres of life."

∽

There is no reason to doubt that all the new *products,* the new comforts, conveniences, things to buy, were in themselves intrinsically compelling, even intoxicating, to the buying public. The heroine of Sinclair Lewis's novel *Main Street,* published in 1920, bemoaned her neighbors' obsessive concern with cheap automobiles, dollar watches, ready-made clothes, Kodaks, and phonographs:

> . . . their conviction, that the end and joyous purpose of living is to ride in flivvers, to make advertising-pictures of dollar watches, and in the twilight to sit talking not of love and courage but of the convenience of safety razors.

Nonetheless, considerable concern arose that consumer demand, if unaided, might not be able to keep the balloon aloft. Thus, several important tools to extend both the reach and the grasp of the market's invisible hand were developed in the twenties and continue to shape merchandising strategies today. Both advertising and planned obsolescence began to pump up demand in order to absorb the inexorably expanding supply.

The essence of advertising was its promotion of both consumption and the ethos of wish fulfillment. Using the tools of psychology and data supplied by market research, during the 1920s the advertising industry moved swiftly from its nineteenth-century emphasis on information to probe the nonrational reasons for buying and to stress the emotional satisfaction provided by new products. How far the industry has come can be seen by reflecting on the fact that in those early days broadcasters saw advertising as a shoddy commercial intrusion and—briefly—banned it from the radio entirely. It is significant that even staunch enthusiasts for commerce like Calvin Coolidge and Herbert Hoover agreed, viewing radio as a sort of public utility that should be above commercial interests.

Since the twenties, however, the advertising industry has closely tracked the temper of the times, evolving in symbiosis with the various media. By the forties radio was competing for advertising revenues and had seized the largest share from newspapers and magazines. Along the way the new pitchmen initiated a number of sales devices—giveaways, trading stamps, contests and premiums—that make buying a game.

A crucial innovation in the late twentieth century, direct-mail advertising, has changed our concept of postal communication at the same time that it has greatly raised our consumption of paper. In the same way that single-use objects reach the trash almost before they have been used, the growth of direct-mail advertising from the 1970s onward has flooded municipal solid waste with printed paper whose message is unread. Beyond contributing vast quantities of waste paper, the direct-mail phenomenon, which accounted for 19 percent of all advertising revenues in 1991, signaled the arrival of ever more ingenious, and invasive, merchandising techniques. By the mid-1980s computer data afforded merchants and advertisers access to consumer profiles, which they used increasingly to pitch their wares to individuals by phone and mail. After 1985, direct-mail marketing promotion grew twice as fast as total advertising, and, at $69 billion, amounted to over half of advertising revenues in 1991. In the process the advertising industry has become a major actor in the U.S. economy, with billings that amounted in 1992 to 4.1 percent of the gross domestic product.

Whether ads really sell the products they are pushing is not at all clear. What they have sold successfully is the image of products that promise easy remedies and perfect function. In its time, television has enormously magnified the power of advertising imagery by wedding it to analogous representations of "the way Americans live" conveyed in the soap operas and sitcoms the advertisers sponsor. In his book on advertising, *The Golden Fleece*, Joseph Seldin constructs a tongue-in-cheek composite of the American lifestyle that emerges from the ads:

> The man of the house rose that morning and brushed the rich, creamy lather into his stubble and got a perfect shave as usual. It left his face soothed and refreshed. . . . At breakfast he greeted his wife with a good-morning kiss as she handed him a glass of juice to help him fight fatigue, colds, and maintain his alkaline balance all day. There was wholesome goodness in every sip. . . . After breakfast he lit a real cigarette. . . . He glanced at his watch, famous for accuracy, and realized he would have to hurry to catch

the 8:15. . . . Meanwhile, back at the ranch-type house, his wife zipped through the breakfast clean-up thanks to the all-purpose cleanser that had twice the cleansing power. . . . The kitchen, with its perfect working surfaces . . . turned work into sheer fun. . . . When the fun was over, she went to shower, pausing long enough to look at her lovelier figure she had gained the pleasant, easy way, without violent exercise, without starvation diets, simply by swallowing a few pills and enjoying all the foods she liked. . . .

From the twenties onward, advertisers' messages were largely aimed at women. Isolated from "the world of work" and presumed to be particularly vulnerable to emotional appeals, housewives formed the front line in the consumer army. During the twenties the suburban exodus swelled, and with it the ideal of the fully equipped private "dream house," implying commitment to the family and separation from community. In Los Angeles, the world's largest mass transit system was eradicated during the 1920s by a coalition of real estate and automobile interests, who argued that it would threaten the "viability of the single-family house." Keeping this house stocked and connected to the rest of the world of course depended on housewives' assuming a new role as chauffeurs—whose mobility would ultimately foster whole new shopping environments.

At the same time, the home economics profession emerged as the main outlet for women's "scientific" impulses. The new efficiency experts found a ready audience for their expertise in the growing array of women's magazines. Under these experts' aegis the art of cooking became "food chemistry," opening the family diet to whole new food substitutes, additives, and mixes, while analysis of housework incorporated time-motion studies, promoting the efficiency that could be achieved with the new appliances.

Although the array of processed foods did lighten the cook's load considerably, rather than shortening women's working day, the conveniences and appliances raised standards for the family's quality of life. Once the children were in school, cleanliness, beauty, and provisioning the household became a housewife's full-time job. The magazines continually reminded women that doing these well demanded "intelligent consumers."

The nature of women's role as chief consumer was described

very candidly in a book published in 1929 for the advertising industry by a well-known women's magazine writer and pundit on home economics, Christine Frederick. While enthusing about women's "special talent for spending," *Selling Mrs. Consumer* offered advice on how to manipulate a creature depicted as not overly intelligent and in need of direction, "whose purchases are exceedingly near her instincts." She was, however, eager to learn, and here advertisers could be helpful indeed. Happily, "the American advertiser has taught the American housewife to think constructively of her job."

Dedicated to Herbert Hoover, *Selling Mrs. Consumer* began with a paean to "consumptionism" as "the greatest idea that America has to give to the world," and traced the source of national economic problems to "the consumer's failure to consume more." People would have to get used to the idea of cutting savings to spend more on new goods, services, or ways of living. As an antidote to the "medieval" habit of thrift, Frederick advocated "*creative waste,*" through which the housewife looks "*to a larger end, beyond the draining of the last bit of utility.*"

Pointing disapprovingly to the European habit of buying products "*to last just as long as possible,*" she cited as a welcome contrast younger American housewives' "disinterest in quality and durability." Although Frederick did not advise her readers in advertising and industry to market shoddy merchandise, she frankly advocated "progressive obsolescence." Housewives love it, happily throwing out of their houses "much that is still useful, even half-new, in order to make room for the newest 'best.'" And making way for the newer and the better "increases the national income." Her audience was unlikely to disagree.

Frederick was only echoing what U.S. industry was discovering, a recognition that has bedeviled capitalist enterprise since that time: the relative finitude of demand in relation to supply. Now that we also have recognized real constraints on the ability of our environment to handle our wastes, that continuing conundrum raises insistent questions.

Looking at the evolution of our consumer economy since Frederick wrote, we cannot fail to wonder whether we can continue moving upward in economic growth without falling back in quality of life. Can we maintain our material life at the present level while also rescuing our natural environment from the stresses consequent upon that lifestyle? As recently as 1966, that question was not evident to even so perceptive a social critic as Vance Packard. In *The Waste Makers,* he dealt trenchantly with the economic, spiritual, and resource drains resulting from what he called "overconsumption," but did not foresee the limited capacity of the biosphere to absorb all that we can produce. Recognizing, and acting upon, that limit will change our lives.

4

~~~~~~~~~~~~

Wealth and Waste

Nature's instructions are always slow, those of men are generally premature.

—JEAN-JACQUES ROUSSEAU, *Emile*

What will the axemen do when they have cut their way from sea to sea?

—HENRY DAVID THOREAU

FOR ABOUT FORTY years, following the Second World War, Americans thought they could have it all, and for that brief period, they were pretty nearly right. What is amazing is how fast it all happened. Technological and institutional innovation spurred the swiftest growth and widest diffusion of wealth in history.

But as often happens when we get what we wish for, the good news was also the bad news. After a short time of unalloyed delight in the fulfillment of our material dreams, anxiety began to dim the joy. Three decades after Rachel Carson, we have increasing reason to fear that production and consumption are blighting natural systems. In this chapter and the next, we will take a long look at the destructive by-products of all that growth—dividing them somewhat artificially into waste (treated in this chapter) and pollution (in the next).

It is doubtful that any people anywhere at any time have been more confident of their ability to control their own fate than Americans during much of the post–World War II era. The industrial process that created the consumption revolution in the twenties and fell into partial eclipse in the thirties rebounded to fight that war and, in the aftermath, nearly succeeded in universalizing the national dream of abundance within American borders. Our dominance after the war was assured by our military might, but the chief spoils of our victory lay in the uniquely unscathed industries pumped up by the war. At the same time in the postwar period, highly visible leaps in technology and science—resulting in radar and faster planes, in televisions and stall showers, in cures for polio and pneumonia, in harvests that increased fourfold—fostered the conviction among Americans that material problems existed chiefly to be solved.

Most crucially, our "invention" of nuclear weapons and energy seemed to rival the power of nature itself. Hiroshima and Nagasaki appeared to overshadow the destructive force of hurricanes, earthquakes, and even volcanic eruptions. Space travel preoccupied Americans from the 1960s onward. Rocketing to the moon the astronauts saw the earth below, round and familiarly contoured, like a classroom globe. Our planet seemed to be getting smaller as our science made us bigger.

As a society we aspired to banish the age-old scourges of indigence and illness—with initial success. By 1960 we had seemingly routed the epidemic dread diseases of smallpox, cholera, tuberculosis, and polio. For a time, from the 1950s to the mid-1970s, we felt ourselves closing in on poverty. Since then we have doubled our gross national product.

As we straddle the 1990s, we're much less confident of our power to alleviate either disease or need, yet in one way our wealth has indeed trickled down. Although the plague and the poor are still very much with us, even the least affluent members of our society have, and throw away, more *things* than their forebears could have dreamed of. New products and processes poured forth so swiftly in the postwar period that Americans came to take rapid technological change for granted. With fertilizer, pesticides, and herbicides, chemical corporations altered the nature of American agriculture. They also delivered a *materials* revolution worldwide. With the creation of plastics and synthetics, they transformed the composition of most manufactured objects. More recently, the electronics industry has played a major role in maintaining consumers' delight in *new things*, while

behind the scenes changing the way everything works. The biotechnology industry, a relative neophyte, now shows signs of rev-
olutionizing science's interaction with human and horticultural
systems, in health care and agriculture.

From the war's end until the mid-seventies, the ability to ac-
quire worldly goods was increasingly broadly bestowed among
Americans. Thus, at least for a time, reality seemed to be converging with the nation's long-held image of itself as the ideal
locus for the pursuit of happiness. After falling steeply during
the Depression from its 1929 high of $103 billion, the nation's
GNP soared during the war, rising from $100 billion in 1940 to
$200 billion in 1944. During the 1950s it jumped again, from
$300 billion in 1950 to $500 billion in 1960. As the rest of the
world recovered from the war, the United States enjoyed an unparalleled monopoly of the world's wealth and productive power.
In 1949, with only 7 percent of the world's population, we had
42 percent of the world's income.

At the time, our per capita income of $1,453 greatly exceeded
that of the other industrialized nations. Canada, New Zealand,
and Switzerland, which had not been hit by the war, were in the
$800–$900 range, Sweden and Great Britain were around $700
to $800; and all the rest fell below that level. In national terms
this meant that much more money was available in the United
States for both public and private projects contributing to economic growth, like research, roads, and education.

The expansion of college education was particularly striking.
The growth of higher education was of course both a cause and
effect of changing income levels. Between 1945 and 1975, enrollments jumped from under 20 percent of the eighteen-to-
twenty-one-year-olds to nearly 34 percent of that age group
(compared with 1.7 percent in 1870).

In a country where most people prefer to characterize themselves as middle class, regardless of their actual income level,* in
the postwar years the middle class expanded to actually include
about half of all Americans: In 1929 eight million households,
or 20 percent of the total, were classified as middle-income, and
drew down about a third of total national earnings; by 1957,

---

*According to various studies, 88 percent, 79 percent, and 75 percent of
Americans placed theselves in the middle class.

twenty-eight million households, representing nearly 50 percent of households and half of national remuneration were in middle-income brackets. Correspondingly, those defined as below the poverty line declined from 33 percent in 1940 to 27 percent in 1950 to 21 percent in 1960 to 11 percent in 1970. (The percentages have risen somewhat since then, starting in the later seventies, reaching 14.5 percent in 1992.) The enormous gain in national discretionary income from the late nineteenth to the later twentieth century is indicated in a comparison between household budgets in 1875, when a typical family used most of its income on food, clothing, and shelter, and 1981, when a middle-income urban family of four spent only slightly more than 50 percent of its budget for these essentials.

At the same time, population grew dramatically. After stabilizing at a bit above replacement levels in the twenties and thirties, national population went from 132 million in 1940 to 152 million in 1950 to 185 million in 1961, a more rapid rate of growth than at any time since the first decade of the century. In 1953, *Fortune* magazine estimated, the number of first children represented a 47 percent increase over the number in 1940, second children a 91 percent increase, third children an 86 percent increase, and fourth children a 61 percent increase. The year 1957 represented an all-time high-water mark for American fertility: 3.7 children per mother. As a result of both prosperity and demography, the number of new households increased even faster than population—by 25 percent between 1947 and 1960.

Most of these new families that could afford to move to the suburbs did so. Between 1950 and 1960 two thirds of the nation's population increase occurred in suburbia. By 1990, 46 percent of the nation's population, or 115 million, lived in the suburbs, compared with 14 percent in 1930.* The availability of new low-interest mortgages following World War II meant that by 1956 for the first time half of Americans owned their own homes. During the fifties Americans also resumed the romance with the automobile that had been strained by the Depression and interrupted by the war. The number of families owning cars rose from 50 percent in 1949 to 70 percent in 1958, at which time nearly

---

*In 1990, 31 percent lived in central cities, and less than 25 percent lived in rural areas.

half of the cars on the road were less than five years old. By 1980 there were almost 120 million cars, or one for every other inhabitant of the United States, and in 1991 cars and light trucks on American roads totaled 167 million. In the suburbs, parking lots competed for land around high schools, and parked vehicles clogged the roads nearby.

Spreading out and moving around in the suburbs and on the road, Americans found novel ways to spend leisure time and additional ways to shop, both fostered by the millions of miles of highways newly paved with federal funds appropriated by the Interstate Highway Act of 1956. Year by year, from gypsy camping in farm fields and along the roadside early in the century, to the free municipal campgrounds of the twenties and the tourist cabins of the thirties, to the motels and proliferating campgrounds of the postwar period, the automobile carried more and more Americans away on vacation. Six million people visited national parks in 1936, forty-five million did so in 1970, and three hundred million in 1991.

In the postwar period, spending on leisure accessories also soared. Trailers, campers, vans, boats, and backyard swimming pools and barbecues have become common paraphernalia, routinely on view in suburban neighborhoods. Sports of choice— from basketball to lacrosse to duck hunting, to fitness programs aimed at various sorts of bodily perfection, including vascular conditioning, muscular development, and comeliness—require both equipment and appropriate attire. All this activity has spawned a multibillion-dollar business purveying new products for more and better leisure, displayed, for example, in the seventy-page L. L. Bean catalog devoted entirely to equipment for hunting in 1993.

Shopping became an increasingly important leisure-time activity, not only for housewives but for Americans of both sexes and all ages. The advent of the enclosed, often glass-roofed shopping mall, combining the ambiance of pleasure garden and amphitheater, has turned the shopping center into the most common locus of communal activity.* Less stylishly, the shopper could en-

---

*By 1987 the number of shopping malls exceeded the number of post offices or secondary schools—in 1992, shopping centers occupied seventeen square feet of U.S. land mass per capita.

joy the same holistic buying experience in ubiquitous giant stores like Wal-Mart, pushing an outsize cart through vast quantities and kinds of goods. Perhaps the ultimate shopping experience, however, can now be found at clusters of warehouse superstores offering acres of goods in a single category—say, sporting equipment or pet supplies.

Electrical and, later, electronic appliances increasingly shaped the consumer experience. According to Ruth Cowan, electric power and appliances as well as indoor plumbing and central heating became available to virtually everyone between 1940 and 1980. In 1940 only 53 percent of households had complete bathrooms, but by 1980 the figure was 97 percent. In 1941, 52 percent of families had washing machines and refrigerators; by 1951, 80 percent had washing machines, and by 1980 access to mechanical refrigeration was well-nigh universal. In 1940 33 percent of households had central heating; by 1980 approximately 80 percent of eighty-seven million dwellings had it.

In the decades since the fifties, two magical inventions, the television and the computer, have dominated consumers' imaginations. (After forty years of television, the heyday of the radio seems so very short: from the end of the thirties, when 46 percent of all families possessed the by then perfected wireless machine, until the advent of TV in the fifties.) At the start of the nineties, 98 percent of American families owned at least one television set; 72 percent also possessed a videocassette player recorder, or VCR, introduced at the start of the seventies. Television has deeply affected our lives, from children's ways of learning to political campaigning. Similarly, the computer, melding the functions of the nineteenth-century typewriter, telegraph, and telephone, has profoundly changed the way we work. The electronic circuitry on which it is based promises to provide future impetus for new tools and comforts—and thus new markets.

By the 1990s, the array of electronic appurtenances considered essential by undergraduates necessitated the rewiring of many college dorms. The president of Haverford College, Tom G. Kessinger, characterized the new high-tech dorm rooms as resembling "the flight deck of the space shuttle." Beyond the basics, computers and audiocassette players, ordinary equipment might include high-speed printers, color televisions and VCRs, compact disc players, answering machines, microwave ovens, electric pop-corn poppers, coffee makers, and small refrigerators. At a time of flat enrollments, many dorms' energy consumption increased 3 to 5 percent annually.

Whenever people stop buying—as they did, for example, when the automobile market became saturated in 1929—panic ensues. In the recession of the late 1950s, President Eisenhower went on television exhorting citizens to spend. Even today, pundits lament meager personal savings rates while at the same time Treasury officials agonize over drops in consumption and heave audible sighs of relief when the buying trajectory heads upward again. Through it all, Christine Frederick would be pleased to see, women have remained the lynchpin of the consumption system.

After the war, American women threw themselves into domesticity with practically fanatical zeal. While producing a bumper crop of babies, they were encouraged to focus their talents exclusively on fostering the comfort and health of their families through the arts of cooking, cleaning, sewing, decorating, and beauty. During the anticommunist tumult of the 1950s, women in their role as consumers were seen as major bulwarks of capitalism. In his 1959 kitchen debate with Premier Nikita Khrushchev, Vice President Richard Nixon lauded the housewife's consumption choices as one of the glories of capitalism:

> To us, the right to choose . . . is the most important thing. We don't have one decision made at the top by one government official. . . . We have many different manufacturers and many different kinds of washing machines so that housewives have a choice.

Betty Friedan's depiction in *The Feminine Mystique* of the ideology that ruled women's lives in the late forties and fifties sounds a lot like Frederick's portrait of Mrs. Consumer. Friedan's quotation from Adlai Stevenson's 1955 commencement address at Smith College sums up prevailing attitudes effectively. Speaking about the "crises of the age," the era's preeminent liberal told the graduating women their participation in politics was through their role as wives and mothers: "Women, especially educated women, have a unique opportunity to influence us, man and boy."

Scrutinizing the women's magazines that housewives read voraciously and the advertisers that supported these periodicals, Friedan found that the new image seemed to require "increasing emphasis on things: two cars, two TVs, two fireplaces. Whole pages of women's magazines are filled with gargantuan vegetables: beets, cucumbers, green peppers, potatoes, described like a love affair." Reading through a selection from a thousand or so market research studies, she discovered a basic message: If women feel unfulfilled in their lives of domestic toil, make them feel creative, attractive, or effective through *buying*. One report stated how, for the isolated housewife, even a sense of community can be supplied by a product:

> Deeply set in human nature is the need to have a meaningful place in a group that strives for meaningful social goals. Whenever this is lacking, the individual becomes restless. Which explains why, as we talk to people across the nation, over and over again, we hear questions like these: "What does it all mean?" "Where am I going to work?" "Why don't things seem more worth while when we all work so hard and have so darn many things to play with?"
>
> The question is: Can your product fill this gap?

While this view of women's role held sway, the number of women working outside the home increased, but the material structure of domestic life changed little. The percentage of the female labor force involved in work outside the home rose from 28 percent in 1950 to 32 percent in 1960, continuing upward to 36 percent in 1970, 42 percent in 1980, and 57 percent in 1991. By 1988, 65 percent of women with children under six were in the work force, and dual paychecks were accounting for an increasing portion of the rise in family incomes and spending. So far, however, women's work outside the home has not led to really far-reaching changes in where people live and how they handle the cooking, cleaning, laundry, even child-care tasks, on which so many consumption patterns are based.* Meanwhile, the

---

*Convenience foods may seem to be an exception, but, as trends from the twenties onward indicate, the popularity of heavily packaged foods probably does not depend on how many moms are rushing home from the office to stick them in the oven.

profusion of products designed to add luster to domestic life keeps on growing, as an estimated 250,000 new utensils for home and garden, covering nineteen acres at the 1992 National Hardware Show in New York City, powerfully demonstrated. And although the market for appliances "matured" as most households acquired a full complement, turnover continued high: Thirty-two million "white goods"—metal appliances (stoves, refrigerators, etc.)—were discarded in 1990.

But as women take on *careers*, as opposed to *jobs*, in the outside world, their availability for the role of Mrs. Consumer is curtailed. With alternative satisfactions or preoccupations, buying is unlikely to have the same meaning in their lives. After an upward surge in the 1980s to a high point of 2.1 births per woman in 1990, fertility rates fell from 1991 to 1993, and the probable cause—declining confidence in the U.S. economy—may also give merchandisers some pause.

An apparently protracted economic downturn, with stubbornly high levels of unemployment, has been accompanied by a trend away from the postwar diffusion of wealth. During the 1980s and '90s the poor and middle classes have lost considerable ground to the wealthy. The 1980s saw the sharpest redistribution of wealth since the Great Depression, as the net worth of the top 1 percent jumped from less than 20 percent of total private net worth in 1979 to 36 percent in 1989.

To the extent that these shifts in demographic and income distribution patterns endure, they will certainly affect the shape of the American economy. Whatever the reason, per capita shopping time in malls reportedly has fallen—from an estimated twelve hours per week in 1980 to about four in 1990. Nonetheless, predictions of a new era of austerity and thrift appear premature. The Bloomington, Minnesota, Mall of America, a shopping emporium equivalent to four ordinary malls, was still drawing crowds of dedicated shoppers by Christmas of 1993, sixteen months after opening. After rising from their all-time low of 2.3 percent of annual incomes in 1987 to about 5 percent in 1990, personal savings fell again to under 4 percent in 1993. If "housewives" as a group can no longer be relied on to sustain spending levels, market researchers have not been laggard in seizing on the implications. Perhaps for this reason, more advertising is now being targeted at particular consumer interest groups—in the general media and in magazines and events tailored to, say, joggers, teenage rock fans, and computer hackers.

As the high school parking problem and dormitory electrical deficits indicate, the buying power of the youthful consumer has become a formidable force. Merchandising and advertising efforts now focus intensively on servicing it. While the cost of toys and gadgets was falling and family incomes were rising, even children in moderate-income middle-class families became used to having their own array of creature comforts that had formerly served the whole family, items such as telephones, televisions, and stereos. When students reach college age, credit card companies solicit them heavily, in many instances waiving requirements for credit histories, income requirements, or parental signatures. Thus encouraged, young people who are not yet wage earners have become accustomed to acquiring a vast collection of gear and continually updating it. The booming $12.1 billion yearly wholesale market for high-tech sneakers, growing by 19 percent a year from the mid-eighties onward, is a tangible example of their impact.

To a great extent, the phenomenal postwar growth in consumption gained much of its impetus from science-based research. Even before the First World War, major corporations like Du Pont, G.E., and AT&T had put laboratories to work on invention and improvement of products. With the addition of major government investment in basic research as well as defense-related development that started during World War II and continued thereafter, the period saw a veritable explosion of knowledge-based industries. Preeminent among them were the petrochemical giants.

More than to any other single factor, the consumption boom of the last fifty years can be ascribed to the postwar miracle of petrochemicals. With the demand it created for aviation fuels and synthetic rubber, the war itself forced into mass production technologies that had been evolving in scattered labs and plants in the United States and Europe throughout the twenties and thirties. Cut off from the rubber plantations of Southeast Asia in early 1942, and with a reserve rubber supply equal to only two thirds of peacetime consumption for a single year, the U.S. government called on an industry consortium led by Standard Oil of New Jersey and Dow Chemical to perfect synthetic rubber. They succeeded so well that at war's end synthetics had largely replaced natural rubber in the United States. Simultaneously, oil companies were developing new cracking processes for refining high-octane aviation fuel, raising output from forty thousand barrels a day in early 1942 to over nine hundred thousand barrels a day by V-J Day.

The war changed everything. Less dramatic but perhaps more important than the great military victory were the triumphant technologies that made the victory possible. After the troops came home the immense capacity of the new American petrochemical industry was loosed on the domestic economy. The chemical industry, which until the war had relied on raw materials derived from coal tar, now was inundated by an array of oil-derived feedstocks from refineries built to supply wartime rubber and fuel production. Dow, Monsanto, Du Pont, Union Carbide, and Goodyear all sought new uses for plants built on wartime government contracts.

Despite fears of postwar contraction, industry adapted rapidly. After a decade of depression, followed by fat paychecks and meagerly rationed consumer goods during the war, Americans were ready for boom times. Peter H. Spitz points out in his study *Petrochemicals: The Rise of an Industry* how production for the military accustomed people to new products made of new materials. Both servicemen and women factory workers who had marveled at nylon parachutes, Butyl rubber liferafts, Plexiglas airplane windows, and plastic raincoats were eager to incorporate "modern" commodities into their civilian lives. Demand for durable goods escalated sharply, with production in the fall of 1946 running at 214 percent over the 1935–39 average. Again in 1951–52, with the start of the Korean War, the U.S. refining and chemical industry undertook its largest expansion yet, upping capital spending by 118 percent. Throughout this period, the market for plastics other synthetics exploded—and supply more than kept pace. In the four decades following the war, annual U.S. production of synthetic organic chemicals increased fifteenfold.

Before the war, less flexible and durable plastics had typically been used for curios and toys, or small objects like combs, toothbrush handles, doorknobs, and dinnerware. Spitz points out that in 1942 small bits of plastic might have been used in an automobile for knobs, door trim, brake bands, horn buttons, and safety glass; by 1947, however, manufacturers were installing it liberally, in vinyl upholstery, reinforced polyester fenders, plastic convertible tops, and a variety of plastic finishes. By 1984 the proportion by weight of plastic in an automobile had risen to 6.4 percent (or about two hundred pounds).

In the 1967 film *The Graduate,* Dustin Hoffman's parents' friend told him, "I just want to say one word to you, just one word: plastics." By that time, the revolution in materials had already been in progress for several decades, but much more was still to come. By the 1980s the U.S. economy was producing and consuming more plastic than steel, aluminum, and copper combined. Beyond its wholesale takeover of items in daily use, plastic also played a fundamental role in altering the composition of consumer wastes.

Plastic's economy, lightness, and extraordinary flexibility did much to change the nature of containers and packaging. The once common reuse of bottles, bags, kegs, and bins seemed pointless, and as distribution systems adapted to throwaways, reuse became impossible. Reusable beverage bottles, well-nigh universal in 1950, had pretty nearly disappeared in most places by the late 1970s, as refundable glass gave way to no-deposit glass, metal, and plastic throwaways. Further, package design and convenience were becoming key elements in selling: The layers of paper and cardboard, foil, and plastic were one with the product.

Plastic also made possible the innovations in the containerization of fast food. Taking the advent of McDonald's as an appropriate starting point, from the mid-1950s onward the two industries took off together. Joined in the 1970s and 1980s by a vast panoply of takeout fare available in gourmet shops, supermarkets, and corner delicatessens, they added a monumental increment of single-use paper and plastics to the nation's municipal solid waste.

As a result, the portion of packaging in American household trash exploded in the 1960s and reached one third of the entire municipal solid-waste stream by 1990. Nearly half of this, 40 percent by weight,* was paper, including newspaper and cardboard.

---

*Unless otherwise specified, figures here for percentages of the waste stream are by weight.

Plastics constituted 13 percent of packaging and about 9 percent of the waste stream (20 percent by volume). Glass amounted to 25 percent of packaging and 7 percent of the waste stream, while metals were 9 percent of the packaging and total waste streams. Encased in all that packaging, by 1988, food made up only 7.4 percent of household garbage.*

Beyond their volume, the durability of plastics and other synthetics also qualitatively changed the nature of our litter problem. The properties of plastic constitute a paradox. Although many synthetics cost so little to make that they are intended to be used briefly, or only once, they last virtually forever. As single-use items exploded, communal trash acquired a different scale and character. Seemingly everywhere—afloat in city gutters and caught in remote woods and sands—plastics never decay or entirely disappear.

Detergent bottles, tampon applicators, Styrofoam cups, vast abandoned plastic driftnets abandoned by fishing boats, and up to fifty thousand plastic particles per square kilometer bob gently out in mid-Pacific; umbrellas, salad containers, and Evian bottles blow through San Francisco's Union Square plaza; butane lighters, soda cans, and Baggies scatter across the high slopes of Mont Blanc above Chamonix in the French Alps. During Earth Day festivities in New York City, celebrants left one hundred tons of debris in Central Park. Over three billion single-serving juice packages are sold in one year in the United States, creating thirty-five thousand to eighty thousand tons of waste.

In addition to the packaging industry, the petrochemical revolution also transformed American agriculture. After the war, the oil boom spurred a long-delayed industrial revolution in farming. Relying heavily on petroleum-based fertilizers and pesticides for mass production of single crops, U.S. farms more than doubled their output between 1950 and 1980. They also loosed a flow of chemicals into rural water systems.

Today, petrochemically based materials so dominate the fabrication of ordinary objects that most people probably do not know which of the things they wear, ride, watch, work on, listen and talk to, play with, or inhabit every day are made of, fueled

---

*The other sizable category was yard waste, at 20 percent by weight and 10 percent by volume.

by, cleaned or maintained with synthetics and which came from "natural" materials. At the same time, technological innovation has made repair and reuse increasingly expensive compared to buying new things. The new materials spun from petrochemicals are so cheap that mending seems merely a compulsive display of thrift. As electronic gadgets have proliferated, the gap between sophisticated devices and ordinary technological expertise has widened progressively. The workings of automobiles and electronic appliances have become encapsulated in "black boxes" that are largely impenetrable to the average consumer with a tool chest, and are frequently designed to be thrown away when they break down.

Garbage we have always had with us. Yet during the past fifty-odd years we have so increased the volume and toxicity of our leavings as to visibly overload the biological systems on which we ultimately depend. While manufacturing wastes and the residues of daily life are usually viewed, and handled, separately, it is ultimately impossible to disentangle the driving power of production and consumption. As the great expansion of productivity reversed the imperatives of supply and demand, the role of household waste as virtually a third dimension of the production/consumption relationship stood fully revealed.

Unfortunately, it has taken us some time to realize the Faustian nature of our bargain. In ancient times people would have anguished about provoking the gods. Today we worry about risks of cancer or genetic damage, about poisoning our water supplies, about toxic wastes permeating the land so deeply we may never be able to cleanse it.

By now, production and consumption have so changed our habitat that fully calculating the impacts might well be impossible. A little more than two decades ago, the near death of the Great Lakes, representing about 20 percent of the world's and 95 percent of the nation's surface freshwater supply, drew public attention to the poisonous power of chemical wastes. Lake Erie strangled on its own algae, and Cleveland's Cuyahoga River caught fire.

After the passage of the U.S. Clean Water Act in 1972 and the investment of $9 billion by the United States and Canada, many of the most visible contaminants, like raw sewage, industrial chemicals, and the phosphorus in laundry detergents, were dissipated. Nonetheless, large amounts of effluents continue to flow into North American waters. For example, at least 56 million pounds of pesticides are deposited in the Great Lakes watershed every year. Those that flow into the Lakes become increasingly concentrated with time because less than 1 percent of their waters flow out of the Lakes every year. The chemicals also build up in the Lakes' food chain. Although the concentrations of PCBs (polychlorinated biphenyls—highly toxic chemicals formerly used in electrical transformers)* in most of the waters of the Great Lakes Basin is below the threshold of measurement, after a tiny water flea is eaten by a mysid, the mysid by an alewife, and the alewife by a gull, the gull's eggs may have a concentration of PCBs twenty-five million times greater than that of the lake waters. The concentration in lake trout and coho salmon of PCBs, many of them airborne and deposited long after the chemicals were banned in this country, is emblematic of the new problems we face.

Even setting aside the issues of atmospheric and air pollution, and concentrating on residues in land and water, the scale of the waste is beyond ordinary human comprehension—so vast as to be almost meaningless. As we will see in the following chapter, it has polluted residential and rural water supplies; agricultural land and water; ports, beaches, and wetlands; river and ocean fisheries; military plants and nuclear reactors; strip and shaft mines; forest and mountain trails.

What we as householders put in garbage pails constitutes only a fraction of our society's waste problem, though for most people, understandably, it symbolizes the garbage crisis. Municipal solid waste, comprising household garbage plus some trash from offices and restaurants, is indeed a growing problem. Just as the

---

*PCBs decompose very slowly over a period of several decades. It was only in 1966 that scientists began to suspect that their accumulation and concentration in fatty tissue could be hazardous to health. Although it has not been proved that PCBs cause cancer in humans, tests on laboratory animals show that they contribute to the development of tumors and are a cause of death. In 1977 the U.S. government banned their production and in 1979 barred their use without a permit.

accelerating ability to transform resources into products shifted the motive power in capitalist economies from production to consumption, our analogous ability to keep more people alive longer means there are more consumers. Though U.S. population has soared, garbage output per person has increased even faster: Since World War I, the American population has grown by 34 percent, while solid waste has shot up 80 percent. With per capita garbage generation rising from 2.7 pounds a day in 1960 to 4.0 pounds in 1992, household waste has grown more than twice as fast as population during the last few decades. The United States generated more than twice as much garbage per capita as Japan and nearly three times as much as West Germany in 1989. Although the 200-some million tons of municipal trash we generate annually account for no more than 3 or 4 percent of our national waste stream, how we cope with it will do much to shape our responses as a society to the vast universe of waste that has underlain our economic and military growth.

Human wastes flushed into the sewage system or backing up from septic fields also pose growing problems. According to a 1989 report by the Natural Resources Defense Council, 5.0 trillion gallons of industrial waste water and 2.3 trillion gallons of sewage in various stages of treatment are released into U.S. coastal waters each year, and 3.6 trillion gallons are released into rivers that flow into the ocean. Not surprisingly, the upsurge in construction of federally subsidized sewage treatment facilities during the 1970s and '80s led to a doubling in the generation of sludge (and sludge volume is predicted to double again by the year 2000). With the termination of federal subsidies in 1991, both our sewage and our sludge problems are growing murkier.

Beyond household wastes, the garbage problem includes mining, forestry, manufacturing industry, farming, and defense. All these generate great volumes of debris and contaminants. Until recently these wastes have been handled for the most part away from public scrutiny.

Both mining and forestry seriously affect the environment. The mining industry, at about 2.7 billion tons a year by far the largest single U.S. waste generator and among its largest energy-users, has stripped away and only partially replaced great swaths of the American landscape. It has also left mammoth piles of tailings from underground mines—fifty billion tons across the nation—that rank among the nation's worst toxic dump sites. Pollution from mining, from both abandoned mines and the runoffs of metals and chemical poisons from processing, affects perhaps 10 percent of waters monitored for diffuse (commonly called "nonpoint-source") pollution. Timber cutting and road building add perhaps another 5 percent.* Disposed of by private owners in relative isolation, the refuse of the two industries is considered here chiefly as a cause of water pollution and an object of remediation through the federally mandated Superfund program for cleaning up toxic waste. Insofar as exploitation of raw materials is central to producing waste, however, debris and contamination from both mining and forestry will be significantly affected by how we deal with the overall waste problem.

Wastes from manufacturing industries were similarly hidden from view before Love Canal in 1978, the dioxin scare at Times Beach, Missouri, in 1983, and other toxic waste scandals highlighted threats to community health from industrial processes. Most large enterprises handle their wastes on the site where production takes place. In order to maintain clean machinery and an orderly workplace they collect residues and by-products for recycling, treatment, storage, or simple dumping. With their reputations to consider, most businesses, it was usually assumed, would manage their refuse in a socially responsible manner, that is, would not permit it to become an eyesore or a public nuisance. Therefore, until a few years ago officials for the most part left industrial waste disposal in the hands of industry.

It is difficult to establish the size of the industrial waste stream with certainty, because it cannot be accounted for in community landfill, incineration, and recycling tallies. According to 1992 EPA estimates, however, nonhazardous industrial waste

---

*Aluminum mining is particularly wasteful of both subsidized electricity and by-products. Every ton of aluminum produced leaves about a ton of corrosive "red mud," composed of metallic oxides and other contaminants; smelting a single pound of aluminum requires six to eight kilowatt-hours of electricity (a typical smelter producing a hundred thousand tons a year consumes as much electricity as a small U.S. city).

amounted to 7.9 billion tons annually, and industrial hazardous waste amounted to 300 million tons. Of these amounts, almost all nonhazardous and 94 percent of hazardous wastes are handled on industrial sites.

Most hazardous and a substantial portion of nonhazardous industrial wastes emanate from chemical and petrochemical processing and metals utilization. According to the EPA some 70,000 chemical compounds have been introduced into industrial use since World War II, and 1,000 to 2,000 new ones emerge every year. U.S. industry produces more than 100 million tons of chemicals annually. At every one of the 1,200 industrial hazardous waste sites so far identified by the EPA as most urgently demanding cleanup contamination involves petrochemicals.

Chemicals also suffuse industrial wastes defined as "nonhazardous" so thoroughly that the distinction is in fact pretty murky. They are found in various direct applications (inorganic chemicals, plastics and resins, petroleum refining, fertilizer and agricultural chemicals, organic chemicals, rubber) that make up more than 20 percent of industrial debris; in pulp and paper residues, which make up more than 30 percent of the total; and in textile- and leather-processing wastes as well. When we consider another EPA estimate, that 9.7 billion pounds of industrial chemicals are legally released into U.S. waters each year, it is evident that the pervasiveness of chemicals makes it very difficult to draw a line between hazardous and nonhazardous wastes or even to clearly delineate what is counted as waste.

Chemical inputs in agriculture have also grown phenomenally since World War II, at first as a miraculous way of accelerating productivity and later as a result of dependency and, to some extent, diminishing returns. When the "miracle" insecticide DDT was invented in 1939, people thought it would permanently rid the world of flies and other insect pests. Its use by farmers grew dramatically in the 1950s. During the Korean War, fertilizer production also shot upward. The U.S. government set a goal of 2.9 million tons of nitrogen a year for 1954, 80 percent higher than the 1951 output. From that point onward, U.S. agricultural methods changed radically, as farmers phased out crop rotations that kept down pests and recycling of animal wastes that supplied nutrients, in favor of heavy infusions of chemical fertilizers, in-

secticides, and, in the 1960s, herbicides.* The industry that arose
to supply these inputs also developed new and higher-yielding
hybrid crop varieties that required massive amounts of chemical
nutrients and pesticides. Agribusiness marketed seeds, fertilizers,
and pesticides to farmers as a package.

As a result, in the thirty years following 1951, fertilizer use in
the United States quintupled, eventually rising to 23.7 million
tons at its high-water mark in 1981. At roughly the same time,
the production of pesticides, some fifty thousand formulations,
increased tenfold, rising from virtually zero at the end of the war,
when synthetic pesticides were introduced, to 1.1 billion pounds
a year in 1993. Americans grooming their lawns consume tons
of fertilizer and 70 million pounds of pesticides annually—that
is, four to eight times as much per acre as farmers, and 6 percent
of the national pesticide and fertilizer total. Three times as much
pesticide is used today as in 1962, when Rachel Carson first
sounded the alarm about its toxicity. (Much of what is sprayed,
perhaps as much as 85 percent, never reaches the plants but
drifts away, sometimes as far as twenty miles.)

The action of these chemicals reinforces the need steadily to
increase their use, as they create a dependency that is analogous
to addiction. Pests and weeds grow resistant to a particular pesticide and consequently ever heavier applications are needed. In
1948, American farmers used 15 million pounds of insecticides
and lost 7 percent of their crops to insects; today they use 125
million pounds and lose 13 percent. As with pesticides, the effectiveness of fertilizer also diminishes. Whereas in the 1960s an
additional ton of fertilizer applied to corn fields multiplied yields
by as much as twenty tons, today an additional ton of fertilizer
may increase harvests by only a few more tons.

After the early 1980s some sense of diminishing returns set in,
but government policies and farmers' behavior changed very
slowly. With evidence of declining effectiveness and toxic
buildup, by 1991 fertilizer use had fallen from its 1981 high to
18.5 million tons. Yet U.S. government policy continued to sup-

---

*The most controversial use of herbicides was in the defoliation campaigns
of the Vietnam War, during the 1960s and early 1970s. In the aftermath, Vietnam veterans sued the federal government for illnesses and genetic damage
allegedly resulting from their wartime exposure to the herbicide Agent Orange.

port high input monocrop agriculture, which heightens dependence on fertilizer and pesticides. The system makes it very difficult to switch crops or to use alternative methods like crop rotation to reduce erosion. Thus the rich topsoil that is one of our nation's greatest resources is also one of our greatest sources of polluting waste. Eroded soil sediments now account for about half of contaminants in streams, rivers, and lakes.

The remaining major source of toxic waste and related human and environmental ills, nuclear and military residues, reportedly accounts for the largest quantity of hazardous waste. Nonnuclear hazardous wastes have amounted to more than five hundred thousand tons each year. Atomic wastes, though especially difficult to deal with, are much smaller in volume. There are reportedly, more than seventeen thousand toxic waste sites located at domestic military installations.

Like our other waste problems, the uniquely intractable dilemmas posed by nuclear residues stem from our postwar technological confidence and ignorance. The beginning of the atomic age, with Enrico Fermi's creation of the first atomic pile under the stands of the University of Chicago stadium in 1942, was a heady time for science. Extraordinary intellectual achievements went hand in hand with a high sense of purpose. Risk was always implicit in the nuclear endeavor, in its inherent destructive power, and in the presence of radioactivity. Among the many misgivings expressed by scientists, however, little concern was directed at radioactive waste. If one could split the atom, surely one should not have to worry about finding a place to put the debris.

The debris and contamination created by the multibillion-dollar U.S. nuclear weapons industry probably would not have been included in this waste catalog a few years ago. Until the end of 1988, the Department of Energy kept the vast network of weapons plants, sprawling over thousands of square miles on federal reservations in thirteen states, isolated from scrutiny. From what we know now, however, it seems that the Cold War ended just in time. U.S. nuclear weapons complexes are in very poor shape.

That the Savannah River (South Carolina) and Hanford

(Washington) complexes would end up this way could not have been imagined by their founders, who built on the heroic scale required by their technologies. In Pat Conroy's *Prince of Tides,* a fictional account of how part of a Southern seacoast county is denuded of its population to make way for a reservation devoted to nuclear weapons production, fiction seems no more outlandish than fact. Irving S. Shapiro, former chief executive officer of Du Pont, tells how they actually did it: "You mobilize all the guys; you go down and wipe out a couple of towns in South Carolina. You go down to visit and you're overwhelmed by the scale of it, but you create it and make it work." Du Pont's Aiken, South Carolina, weapons plant, at Savannah River, extended over three hundred square miles (only half the size of the prototype at the Hanford reservation in Richland, Washington). At Savannah River more than fifty underground tanks, each as big as the dome on the U.S. Capitol, house the spent fuel; weapons-grade plutonium was chemically separated in huge concrete rooms known as "canyons." At the Site X weapons complex in eastern Tennessee, the equivalent of the annual timber output of Minnesota was used during World War II to build what was then the largest roofed structure in the world.

The waste problems of the civilian nuclear power industry, though separate, are causally and culturally related to those of the postwar weapons industry, stemming as well from the technological innocence of those early days, when we didn't know enough to fear what we didn't know. Fifty years later our dimming zeal for atomic energy is directly related to our failure to find safe places for the debris. And we have accumulated a lot of it. Nationwide, as many as forty-five thousand sites may be polluted in varying degrees with radioactive materials.* As we entered the 1990s, that waste and the collective national resistance to the disposal methods proposed so far blocked the development of nuclear power in the United States. Led by California in 1976, many states have passed laws tying the development of commercial nuclear power to solving waste problems.

---

*This estimate includes medical and manufacturing wastes as well as those of military and energy industries; about twenty thousand of the sites, and many of the most contaminated, are owned by federal agencies, particularly the Department of Defense and the Department of Energy.

Current government plans for the disposal of low- and high-level nuclear wastes each raise their own technical and political problems. Low-level waste consists principally of contaminated materials from nuclear power plants and diagnostic medical services. While most of it can be expected to remain radioactive for around a hundred years, it can also contain intensely radioactive materials that would remain "hot" for a much longer time. Since 1978 all low-level radioactive waste has been buried in three states, Washington, South Carolina, and Nevada, but for more than ten years Congress has been fruitlessly trying to get the other states to take responsibility for the waste generated within their borders.*

The problems with low-level materials look easy, however, compared with those of siting a repository for high-level wastes. These materials consist principally of highly radioactive spent fuels, about 95 percent from commercial power plants and the rest from weapons manufacture. These wastes are extremely lethal—technicians at one facility joke that the spent fuel is " 'self-protecting,' meaning you'd die before you could get close enough to steal it"—and will remain radioactive for ten thousand years or longer. Despite the halting growth of the nuclear industry worldwide, the volume of high-level wastes from plants built in the past several decades is growing fast; it is now three times that of 1980 and twenty times that of 1970.

High-level wastes remain aboveground (in the United States more than twenty-two thousand metric tons are stored in glowing blue cooling ponds adjacent to reactors), while governments around the globe seek suitable locations for permanent underground burial. None has yet been found. The problem, as a 1990 National Research Council report pointed out, is that the more we study it the more we increase "the number of ways in which we know that we are uncertain." In 1987, Congress designated Yucca Mountain, Nevada, as the sole U.S. site, over the vehement objections of Nevadans, after Eastern states forced the cancellation of the as yet unsited Eastern repository in 1986. Critics of

---

*A federal mandate to locate regional sites for low-level waste disposal facilities (initially supposed to happen in 1986 and then in January 1993) has turned into an occasionally farcical game of NIMBY on a national scale: Within their regional groupings the states have spent most of their energies passing the buck.

the plan point to numerous seismic faults at the Nevada site, which is twelve miles from an extinct volcano.

An important aspect of our civilian nuclear waste problem stems from the short life expectancy of reactors (between ten and thirty years). Eventually every reactor will have to be decontaminated and dismantled. Because a large commercial reactor has never been decommissioned, we are just beginning to grasp the extent of the problem. It is estimated that the seventy or so U.S. plants that will have to shut down by the end of this century may cost more to decommission than they did to build—from $175 million to $750 million each. And the dismembered parts will swell the nuclear waste stream. Environmental analyst Cynthia Pollock Shea estimates that the contaminated debris from one large decommissioned reactor would equal one fourth of all low-level waste disposed of in the United States each year.

As the 1980s drew to a close, these supposedly discrete problems of household garbage, sewage, industrial toxics, agricultural chemicals, and nuclear residues began to blur. It seemed you couldn't turn around without running into a waste problem. Cumulatively, all this waste generated by modern economic life was undermining the material world that sustains us and runs our economy.

5

Swamped

Nature has poured forth all things for the common use of all men.

— AMBROSE, Bishop of Milan

As crude a weapon as the cave man's club, the chemical barrage has been hurled against the fabric of life.

— RACHEL CARSON, *The Silent Spring*

SUDDENLY EARTH, THIS vast planet, seems very small. Our production of people and goods has accelerated so quickly in the past fifty years that we are rapidly filling up a planet that earlier societies considered almost infinite in extent. The residues of all this growth, which once posed chiefly management problems, now loom so large that they threaten the carrying capacity of air, land, and water. Because we keep on growing, our real efforts at pollution control have not succeeded in reversing the rising tide.

Given the magnitude of our cumulative wastes and the limits of the scientific analysis devoted to studying them, we are far from understanding their impact. We do know that all wastes— household, industrial, agricultural, and nuclear—in some way change the air, land, or water where they are dumped. What we call "pollution" runs from dirty to poisonous, the degree of contamination being determined by the effect it has where it falls.

Experts who rank current threats to our environment most often put damage to the atmosphere first, followed by loss of species and habitats. Because so much is incomplete, in both our understanding of impacts and our programs for remediation, it seems particularly important to ponder what we don't know and how far our efforts to cope have fallen short. A growing backlash against massive spending to clean up toxic waste sites reflects confusion about risk as well as the policies we have employed to reduce it. As we look at how wastes are contaminating our land and waters, we will not try to assess relative risks or money costs, but simply consider the volume of materials entering the ecosystem as a result of human activity and reflect on its visible impact.

Our problems in understanding what we are up against emerge clearly when we grasp the extent and importance of small-scale and diffuse "nonpoint" sources of pollution. Until recently, most attention and regulation was devoted to major "end-of-pipe," or "point" emitters of pollution, those whose emissions from stacks or drains could be relatively easily identified, measured, and monitored. Currently, however, the effects of more mundane and diffuse emitters are becoming an increasing focus of concern.*

The EPA defines nonpoint-source water pollution as "diffuse pollution resulting from land runoff, precipitation, atmospheric deposition, drainage, or seepage, rather than a pollutant discharge from a specific, single location." Whereas main point sources would include heavy industries and utilities, nonpoint emissions come from individuals and private homes; farms; small commercial or manufacturing businesses such as dry cleaners, service stations, or metalworks; do-it-yourself oil changers; leachate from dumps; and road runoff (oil, asbestos).

Because of their elusive quality, these nonpoint sources are emerging as the most intractable U.S. water-quality problem. While some progress has been made in curbing point-source emissions from industry, transport, and sewage, nonpoint pollution flowing from the daily activity of millions of individuals who cannot be required to get permits or to install "end-of-pipe" emission controls, is not easy to regulate away. A farmer whose fertilizer and cow manure wash into a stream running through

---

*EPA efforts are still devoted overwhelmingly toward "point" sources; states have been able to allocate a few million dollars at best toward solving billion-dollar problems.

his fields, or a jeweler who flushes his used solvents down the toilet, makes a very elusive target for regulations. So do motorists whose oil or transmission fluid drips along the highway and homeowners whose septic fields back into nearby aquifers.

For diverse reasons, estimates of the impact of cumulative non-point flows must be less than exact. Pollution levels have been measured for only 20 percent of all coastal waters, 29 percent of stream miles, and 41 percent of lakes and reservoirs, and for a much smaller portion of groundwaters for which no federal criteria currently exist. Nonetheless, the EPA estimates that "76 percent of impaired acres of lake water, 65 percent of impaired stream miles, and 45 percent of impaired estuarine square miles are affected by nonpoint source pollution." Further, a Natural Resources Defense Council (NRDC) study estimates that in 1988, 70 percent of all sediment, 90 percent of fecal and other coliforms, 80 percent of nitrogen pollution, and 50 percent of phosphorus pollution came from diffuse sources. Oil and other chemical fallout from daily urban life are also flushed by rain from storm sewers, gutters, streets, and parking lots, as well as industrial sites and municipal sewage systems into some 5 percent to 15 percent of the waters that have been assessed. In addition, land development, construction, and highway building deposit sediments and toxic runoff in an estimated 1 percent to 5 percent of our surface waters.

Farm chemicals and soil erosion account for a major share of diffuse water pollution nationwide. Prior to 1979, it was believed that pesticides did not leach into groundwater, but in that year, dibromochloropropane or DBCP, a fumigant used on cotton, soybeans, fruits, and vegetables, was found in California wells, and aldicarb, an insecticide used to kill roundworms, was found to have contaminated wells on Long Island. By 1985, monitoring revealed pesticide contamination in over half the states. According to the EPA, 50 percent to 70 percent of U.S. waters monitored for pollution are adversely affected by deposits of eroded soil or fertilizer and pesticide runoffs.

Cleaning up this pollution would be highly expensive. If the affected counties could afford it, monitoring contaminated water supplies would cost nearly $1.5 billion a year, and hooking households with contaminated wells to public waterworks would run about $12,000 per household. Cleaning up contaminated aquifers by injecting biological or chemical agents to detoxify contaminants, or pumping groundwater to the surface for treatment and then returning it to the aquifer, is highly costly and in some cases beyond the reach of current technology.

Though more easily identified and regulated, point sources of pollution also represent a vast and costly problem. The largest specific sources are nuclear and military plants. Because of the dangers involved and the need for secrecy, U.S. weapons plants ordinarily are located in sparsely populated rural landscapes. For the past four decades the military has operated its munitions installations with virtually no environmental constraints, releasing thousands of tons of radioactive materials into the environment during that time. Thus, like farm chemicals, toxic and nuclear by-products from these plants leak into nearby rivers and streams, sometimes affecting underground water as well.

The military investment underlying U.S. superpower status has left a unique toxic waste problem paralleled only by the extraordinarily grave contamination of the former Soviet Union. At the largest and oldest military site, the weapons plant operated for the government by Westinghouse in Hanford, Washington, nuclear waste dumped into the ground is seeping into the Columbia River and spreading through about twenty acres of soil and groundwater a year. In addition, 149-million-gallon storage tanks, including perhaps 20 tanks containing potentially explosive trapped hydrogen, are cracked or leaking. At the Savannah River weapons plant, operated until 1989 by Du Pont, problems are similar, with leaky storage tanks and steady seepage of radioactive and hazardous wastes from enormous ponds into the Savannah River and nearby swamps. More than 250 waste dump sites dot the reservation.

Three other major nuclear weapons sites—the Oak Ridge National Laboratory, America's first nuclear weapons plant, in Tennessee; Ohio's Feed Materials Production Center in Fernald; and Colorado's Rocky Flats complex—are now most notable as key toxic waste problems. Raided in June 1989 by FBI agents looking for deliberate violations of environmental laws, the Rocky Flats plant, run by Rockwell International, is perhaps the most notorious. The complex and its environs, sixteen miles from the center of Denver, is regarded as one of the most contaminated places on earth. Over the last forty-odd years the installation has been subject to plutonium fires, accidental releases of perhaps two thousand curies of tritium (a radioactive gas that is the main ingredient of hydrogen bombs—a reactor core may have ten billion curies), three thousand barrels of leaking plutonium-contaminated oil, and cumulative releases of plutonium that have raised contamination in suburban Denver soils to levels ten times normal. The DOE estimates that seven nuclear bombs' worth of plutonium escaped into the air ducts of the plant over its thirty-eight years of operation. Occasional winds at Rocky Flats blow plutonium lodged in its soils over downtown Denver, causing scientists to call the area a "creeping Chernobyl."

Beyond the billions of tons of existing radioactive contamination already described, industrial processes and by-products involved in the manufacture, maintenance, and repair of weapons, aircraft, ships, or trucks at various military bases throughout the country have generated hundreds of thousands of tons of hazardous waste each year. The production of ammunition has "mined" thousands of acres with unexploded shells, and the by-products from toxicological-, biological-, and lethal chemical-warfare materials, as well as solvents, paints, and chemicals, sit in thousands of ponds and impoundments. According to a Pentagon report, two thirds of the sites they investigated on air force, army, and navy bases need restoration work.

The estimated cleanup costs jump each time new nuclear cesspools or chemical sumps come to light. Before public attention was drawn to this contamination, the military's calculations centered on the billions of dollars in savings they expected to gain from turning their bases over to private companies or communities. When, in 1988, they realized they were landlords of some *very* hot properties, the DOE estimate of a $91 billion to $128 billion cleanup bill seemed astronomical. Nowadays, calculations run from $100 billion to $200 billion for Pentagon sites and are around $300 billion for DOE nuclear weapons cleanups.

⁓

As enormous as the military waste problem looms, it still appears uniquely containable compared with wastes from civilian production and consumption. The superpower struggle that spawned the military residues is over. Whether or not the vast areas contaminated by the weapons complexes are fully restored, the processes causing that contamination are being curtailed. The same cannot be said of our civilian waste streams, however. Even as we move into an era of heightened environmental awareness, our production and consumption continue to diffuse huge volumes of chemicals and metals.

Pollution from runoff and disposal of consumption wastes emanates from both nonpoint and point sources, including waste treatment plants, landfills, corroded oil tanks, septic fields, lawns, and construction. Two decades after the passage of the Clean Water Act, over a third of the toxics entering our waters still flow from waste treatment plants. Fecal coliform has shut down oyster beds in Louisiana, ended shellfishing in Washington's Puget Sound, and cut into shellfish harvests along the Massachusetts coast and in the Chesapeake Bay. In the summer of 1989, Asbury Park, New Jersey, was fined $1 million after overflow from its sewage plant closed Monmouth County beaches. As sewage washed up along twenty miles of San Diego's beaches in the spring of 1992, the city hustled to mend leaky thirty-year-old pipes carrying sewage offshore and debated whether to upgrade a system designed for 250,000 people and now struggling with the wastes of 1.7 million. By 1995, New York City, long boastful of water called "champagne out of the tap," will probably have to filter its supply from upstate reservoirs contaminated by sewage plant discharges.

Even legally disposed garbage is a source of pollution. In landfills, natural anaerobic (oxygen-deprived) decaying processes often convert benign waste into toxic chemicals, which may pollute the air and leak into adjacent ground and surface waters. With coal mines, landfills account for the largest share of methane emissions. Wayward methane equivalent to the carbon dioxide from ten thousand automobile exhausts is emitted into the air from U.S. landfills annually, as are about two hundred thousand

pounds of volatile organic chemicals—landfill methane is estimated to account for more than 6 percent of methane emissions
globally. According to research done at the University of Southern California, 46 percent of the country's existing municipal
landfills (most, like New York City's Fresh Kills, unlined) lie
within a mile of drinking-water wells, and an EPA study estimates
that nearly 90 percent of municipal landfills pollute nearby
groundwater, while 60 percent of those monitored are tainting
surface waters. More than 20 percent of those toxic waste sites
identified in 1989 as national problems (on the EPA's national
priority list)—including eight of the twenty worst sites—were municipal solid-waste landfills.

Over the past decade, state legislation, and now federal RCRA
(Resource Conservation and Recovery Act) regulations, promulgated in October 1991, have mandated liners, leachate control
systems, groundwater monitoring, and methane collection not
found in most existing landfills: As a result of these technologies,
new landfills have substantially curbed contamination. Yet the
security provided by these systems can only be considered temporary. The durability of various sorts of liners (constructed from
synthetic and clay materials) is unknown, and all will eventually
deteriorate. Thus the polluting impact of landfilling garbage is
only delayed, not ended.

Incineration presents more significant immediate threats to
local air than to water quality, but it, too, poses longer-term
hazards to water supplies through its ash. Much more than
landfilling, incineration chemically transforms large portions
of the waste stream, thereby concentrating toxics—particularly
heavy metals—in the ash that subsequently must be disposed
of in landfills, as well as emitting them through smokestacks.
Air and ash pollution are inextricably linked in the combustion process. Filtering out heavy metals, and sometimes chemicals, from the air only raises their levels in the ash—and vice
versa. Though fossil-fuel combustion and smelting create more
overall metals emissions than incineration, in 1989 incineration was the largest single source of U.S. lead emissions and
the second largest source of mercury emissions. Scientists are
only beginning to study greenhouse gas emissions from incineration, but evidence indicates that incineration of paper with
plastics and other organic materials produces twice as much

carbon dioxide as coal combustion per unit of energy produced.*

❧

Turning to the materials that fuel our consumption, petro-chemicals account for the greatest share of toxics emanating from production processes. Their sources are not always easy to pinpoint. In the first place, oil itself pours unbidden from myriad sources. Twelve weeks after the March 1989 rupture of the *Exxon Valdez* poured almost eleven million tons of North Slope crude into the pristine wilderness area of Prince William Sound, three more tankers spilled their cargo in busy port areas in the space of twelve hours: in Narragansett Bay, Rhode Island, on June 24; in La Porte, Texas, on the same day; and in Claymont, Delaware, on June 25.

Four years after the *Valdez* grounding, most of the beaches around Prince William Sound look normal, yet in 1991 the EPA judged that only 8 percent of the spilled oil had been removed. In 1993, scientists found that much of the damage measured apparently stemmed from an earlier spill that had gone unde-tected. Together the spills have altered the bay's environment severely. Over many miles of shore, oil has sunk deeply into the sand beneath the gravel cover, there to stay, draining in and out with tides, presumably for many years to come. A number of experts contend that some of the worst damage will not appear for years, as the effects of the spill work their way through spawning cycles of fish and breeding cycles of seals, sea otters, whales, and many species of birds. (The northern climes are at least spared the recapitulating effects of oil spills on tropical marine systems, where damage may reverberate for a century or more, as oil trapped by sediments pollutes coral reefs, sea-grass beds, and mangrove forests again and again during each rainy season.)

Dramatic as the tanker ruptures may be, they represent only a fraction of the oil continuously streaming, flushing, leaking, or

---

*Coal produces 18 percent more carbon emissions than oil. Burning garbage may also raise methane levels because incineration produces sizable carbon monoxide emissions, which deplete the supply of hydroxyl radicals, which break down methane.

oozing into our land and waters from ships, rigs, storage tanks, pipes, and sewers. Every year some five hundred million gallons, more than forty times the size of the *Valdez* spill (and more than one gram for every hundred square meters of the ocean's surface) find their way to the sea. In trying to fathom the significance of the flows enumerated below, it is illuminating to reflect that a single pint of oil spilled in a lake or stream cuts off oxygen over an acre, and one quart will foul the taste of 250,000 gallons of water.

Setting aside catastrophic collisions, tankers' routine maintenance practice of cleaning their tanks by flushing them at sea dumps enormous quantities of oil. In the first six months of 1991, tankers spilled 425,000 tons of oil around the world, seven times as much as flowed from the *Exxon Valdez*. According to the International Tankers Pollution Federation, in 1988, 33 percent of the oil released into the oceans reportedly came from "routine operations."

In the year after the Alaska catastrophe, the Wilderness Society estimated that pipeline failures and leaks in storage tanks poured a total volume of oil about twice that of the *Exxon Valdez* rupture into U.S. land and water. In November 1993, the federal Bureau of Land Management warned that bad management along the 800-mile Alaska Pipeline, which carries 25 percent of the oil produced in the United States, had created an environmental disaster waiting to happen. Where leaks do occur, from tank farms, gasoline storage tanks, and refineries they are hard to measure or even identify, but the anecdotal evidence we have gives real pause. It is estimated that oil spilled from Chevron's El Segundo, California, refinery has created an underground pool of two hundred million gallons, and twenty-eight million gallons have spilled underneath the Tosco Corporation refinery near San Francisco Bay.*

In 1987 Northville Industries of East Setauket, Long Island, discovered that over a decade enough gasoline had leaked through a tiny hole in an eight-inch pipe at its storage terminal to form a thirty-acre lake of gas directly on the top of Long Island's aquifer, its only source of drinking water. Since then the corporation has spent over $10 million on an elaborate system

---

*For comparison, recall the *Valdez* spill's eleven million gallons and note also the Gulf War–related Persian Gulf spill of sixty-four million gallons.

of wells, pumps, and filters, recovering about half of the million or so gallons spilled. Although they expect to spend millions more, dissolved gas that has migrated deep into the water table may be impossible to retrieve.

It can come as a surprise that smaller nonpoint sources appear to contribute the largest cumulative share of total oil pollution. Most crucial, an enormous amount of used oil—about 570 million gallons or nearly one fourth of the 2.3 billion gallons of industrial and automotive lubricants sold every year—is disposed of improperly, put out with household or industrial refuse, dumped into the soil, or discharged into the sewer system. Of this, about 180 million gallons is attributable to do-it-yourself oil changers, only 10 percent of whom reportedly recycle residues. The state of Florida, where 90 percent of the drinking water comes from aquifers lying just six or seven feet under the surface, estimates that seven million gallons of used oil are illegally disposed of every year somewhere in the vicinity of those aquifers.*

The continuous siege by oil of the marshland on the periphery of New York Harbor graphically evokes the damage done by the careless usage many still consider normal. In 1991 ornithologist Katherine Parsons told how the harbor's Arthur Kill marshland was stricken by a massive Exxon pipeline rupture in January 1990, followed by three other spills in the next two months. As the center of oil distribution for the Northeast, New York Harbor was no stranger to pollution, absorbing about 1.5 million gallons a year from "a daily agony of small slicks." Even prior to 1990's four-pronged assault, the herons and the egrets there were "sustained by a food web as corroded and tenuous as the oil industry's own infrastructure." But the Exxon spill ended even that imperfect accommodation. Though the birds returned hopefully to nest in the estuary the spring following the 567,000-gallon spills, few of their chicks survived.

---

*Deciding that this was "the kind of battering no state can take over a long period" (the words of Dale Twachtman, the former secretary of Florida's Environment Bureau), the state launched a major program of public education and grants to counties, cities, and one Indian tribe, resulting in the establishment of two hundred used oil collection sites in the next year and doubling existing capacity.

~

The industrial universe fed by the basic hydrocarbon source of both energy and petrochemicals is both vast and murky. Although regulated by federal law, what happens to all the waste generated by industry remains within the domain of business—literally, as most of it is disposed of on the site where it is generated in pits or lagoons. For the outside world that domain remains largely terra incognita, at least until crisis ensues. In Toms River, New Jersey, for example, in 1985 residents discovered after the fact that many years' seepage from toxic-waste burial pits near Ciba-Geigy plastics and dye factories had been gathering in a toxic lake under the Toms River that would take thirty years of pumping to remove.

Although only about 10 percent of all industrial waste is defined as "hazardous," what that term means is a source of further uncertainty. The EPA defines as "hazardous" those wastes that either (1) exhibit the characteristics of ignitability, corrosivity, reactivity, or toxicity, the latter signifying that they contain at a minimum one of eight specific metals or six organic chemicals; or (2) are among the 460 or so commercial products and production-process wastes the agency specifically lists as hazardous. However, only a minuscule percentage of the chemicals in use has been evaluated by the federal government for toxicity. Thus the spectrum of problem chemicals can be expected to grow steadily. For example, wastes currently defined as "nonhazardous" may contain small amounts of hazardous wastes like arsenic, lead, mercury, and strong acids; contain toxic chemicals in amounts below regulated thresholds; or pose substantial threats to human health and the environment that are currently being studied by the EPA.

Although regulation has come a long way since the first federal legislation in 1976, government oversight is still episodic, even with waste classified as hazardous. "Nonhazardous" trash dumps and treatment facilities are required to operate according to federal standards. However, they are actually subject to only occasional scrutiny by state officials responsible for their regulation.

What seems clear is that current methods of disposal of "nonhazardous" industrial wastes pose significant hazards. Vast quantities of raw or only minimally treated industrial wastes are flushed directly into municipal water treatment systems designed to treat household sewage, not toxic metals and organic chemicals. Pretreatment of these flows is mandated only for some industries, for some toxics, and for the largest treatment facilities. Numerous toxic industrial dischargers like laundries and car washes are exempt from pretreatment standards, and standards for many others like textile mills and various sorts of plastics molders have never been promulgated.

But even if regulations were to embrace all discharges, determining who is putting what into the system and with what effect presents short-handed and politically beleaguered municipal governments with a formidable, perhaps often impossible, task. In the upshot, the EPA estimates that no less than 37 percent of the toxics entering U.S. waters are flushed by industry through municipal treatment facilities. In addition to polluting rivers and estuaries, mixing industrial effluents with sewage transforms municipal sludge from a reclaimable resource into a large volume of hazardous waste.

Apart from direct discharges into sewage systems, the management of "nonhazardous" industrial wastes raises other major questions. An Office of Technology Assessment (OTA) study released in April 1990 reports that some 2,200 of the more than 20,000 nonhazardous industrial landfills and holding ponds dispose of toxics like arsenic for small-quantity generators, and a number handle wastes that are likely to be classified as hazardous in the near future. Few, however, are equipped with effective barriers against groundwater contamination. Most of the industrial waste (about 95 percent) is dumped in surface pits or ponds that mingle liquids and solids. These holes in the ground were in fact designed—in more easygoing days—to leach their contents away. The liquids create strong gravitational pressure, pushing wastes into the soil. Only about 5 percent of these impounding ponds have synthetic liners, and only 9 percent have installed monitoring systems that can detect seepage.

When it has surveyed the small number of facilities that *do* monitor leaks, the EPA has found substantial evidence of leakage. In New Jersey and California, 61 percent of the 112 facilities with groundwater monitoring reported contamination, and in more than half the cases the threat was adjudged "moderate to great." Although the overall threat to groundwater nationwide is estimated still to be in the 1 percent to 2 percent range, given the limits of our current knowledge of toxicity and of monitored groundwater, that seemingly minimal figure should not reassure. In the universe of hazardous wastes, our former situation of complete ignorance was far less stressful than the tip of the iceberg we see now.

Since the enactment in 1976 of RCRA, the basic law governing waste management, successive phases of regulation have done much to expose the intractability of the problem and to define goals for action. Under the 1976 law, all hazardous waste facilities in existence as of November 1980 had to obtain permits certifying their compliance with standards on such matters as groundwater monitoring. It wasn't until 1984, however, that RCRA established deadlines for obtaining permits or ceasing operations and required all facilities, whether they planned to continue operating or to close, to clean up any contamination on their sites.

While these laws have undoubtedly changed the nature of the hazardous waste business, they have done little so far to remediate the most serious existing environmental threats. Of the 4,615 known hazardous waste facilities, by April 1990 only 1,711 (37 percent) had been assessed for possible releases of hazardous materials. Of those, 1,422 (83 percent) were leaking, but cleanup remedies had been "studied, proposed, selected, or implemented" for only 95. Nationwide, 2,282, or nearly half of the facilities, decided to close rather than to comply with the new regulations affecting landfill liners, groundwater protection, and financial responsibility for contamination after closure. By February 1991, however, only 9 percent of those choosing to close had received certification of postclosure monitoring and cleanup as well as financial responsibility.

For the EPA, adequate oversight is a problem of resources and priorities. RCRA's emphasis is on the *management* of hazardous waste, and for agency staff charged with carrying out myriad assessment, permitting, and oversight tasks, awarding permits to new facilities or those that decide to comply with the new regulations has priority over policing those that are bowing out. Nonetheless, if facilities fail to complete closure by covering wastes with an impermeable cap and installing equipment to monitor for liquid runoff, they have not taken the first step toward stabilizing contamination. Because of the EPA neglect of closure, the prevention of further toxic seepage on the most polluted tracts is receiving short shrift. Thus it seems likely now that many of these areas may join the elite category among polluted sites—Superfund sites.

It is interesting that the most polluted land areas in the United States (that we know of) are characterized first of all not as festering sores (as in Most Hazardous or Extremely Toxic) but as bankable projects. Named at its inception in 1980 for the pool of federal government revenues allocated for remediation, the Superfund program was empowered under the Comprehensive Environmental Response, Compensation and Liability Act (CERCLA) to clean up contamination itself and to force polluters to remediate sites or bear the costs of doing so. Originally, Congress appropriated $1.6 billion for the job, to be spent over five years. At the time, it was expected that the fund would be phased out as abandoned waste sites were treated, while the RCRA program would remain to prevent the same sort of pollution at waste sites in the future. More than a decade later, the number of priority sites has risen to 1,275 (from the original 406 in 1983) and may well reach 2,000 by the year 2000, while the cost of cleaning them up is estimated at anywhere from $40 billion to three times that much.

A great deal of money has been spent; litigators have taken a substantial share; and charges of mismanagement and fraud abound. The threat of liability, which has constituted the most powerful pressure against corporate behavior, also distorts community and business planning. Unfortunately, Superfund is likely to remain part of our landscape for some time to come. Between 1983 and the end of 1993 the program deleted just 57 sites from the list, and completed remedial construction and removal work on several hundred more. Sites are being added much faster than they can be cleaned up, however, and there may be thirty thousand potentially as contaminated as those now on the list. Although the program's coffers have been replenished twice, in 1986 and 1990—when the fund was raised to $8.5 billion and then $15.5 billion—by the end of 1992 federal costs had run over $15 billion, of which $7.5 billion came out of the fund and the rest from those responsible for contamination.

Devoting such a large share of our environmental spending to remediating toxic waste is certainly a questionable investment of national resources. Moreover, numerous communities are balking not only at the diversion of scarce revenues but also at the disincentives to industrial investment inherent in the regulation. For this reason, the Clinton administration proposed, in early 1994, to ease requirements for cleaning up toxic waste dumps, according to how they were expected to be used in the future. As later chapters will argue, the Superfund approach has not provided a cost-effective way to curb toxic wastes. Nonetheless, because the impact of these toxics is still to an unknowable extent hidden from us, and what we know is so difficult to measure, these costs provide our best representation so far of how pollution has damaged our land.

Of course, land and water cannot really be considered separately. The parcels of toxic real estate scattered across the landscape could be placed off limits to people. But they cannot be cordoned off from the air and water, the rain and the underground springs that wash through them to the rivers, lakes, and sea. Both industrial runoffs and "everyday" activities involving sewage, farming, construction, damming, and diversion have brought severe disruptions of that water system.

The amount of water on the planet is finite. Since the appearance of animate life, all living beings have used the same water, recycled through their bodies, and then through the earth, sea, and atmosphere. We drink the same water that nourished the dinosaurs. Our global stock of fresh water (about nine thousand cubic kilometers a year) seems like plenty for the earth's present population, but much of it is not where people need it or in a form they can use. Although we can't destroy water, we can render it inaccessible, undrinkable, and inhospitable to life.

Over the last two decades we have tried with some success to reverse earlier damage to surface waters. Lakes choked with algae, flammable rivers, and beaches assailed by red tides spurred the U.S. to pass the Clean Water Act in 1972. Setting national standards, regulating industrial polluters, pouring federal money into state water-quality programs, the program succeeded in reducing point-source contamination by industrial plants and municipal sewage disposal and in bringing a 100 percent improvement in dischargers' compliance with permits between 1972 and 1982. The construction of hundreds of primary and secondary wastewater treatment facilities since the early 1970s brought a 50 percent increase in Americans' access to sewage treatment. Concentrations of some chemicals and heavy metals in sediments and fish in some major U.S. harbors have fallen as a result of banning pesticides like DDT and chlordane as well as PCBs and leaded gasoline. To a significantly greater extent than before 1972, dumping raw wastes into public waters is illegal, and with the banning of phosphorus in household detergents, a number of lakes and rivers have cleansed themselves of the green gunk and brown foam that was smothering life within them. Though by no means pristine, New York's Hudson River is again inhabited by some hardy swimmers, and the quantities of PCBs in the river's striped bass have fallen almost to a level considered safe for human consumption.

Most of the progress occurred before the end of the 1980s, however, following a $100 billion investment in wastewater treatment, and we have lost ground since then. Although regulations have led to a moderate, general reduction in pollution from industries and sewage plants, this favorable trend is now seriously threatened by the increase in nonpoint sources of contamination. Rising population and development—the cumulative effects of individuals leading their lives, fertilizing their new lawns, using their cesspools, driving and maintaining their vehicles, and building homes and offices—is overwhelming past progress.

As we have seen, the sewage problem is both "point" and "nonpoint," direct and diffuse. In the contemporary version of the backyard privy, private septic fields rely on the microbial powers of the earth to purify injected wastes. When the demands of the system are manageable this does work. As fresh water runs off the land on its way to the ocean, the wastes it picks up can be degraded by microbes in earth and water if the population density of the cachement area is not too high. If the volume of wastes exceeds the self-purifying capacity of the cachement area, however, the wastes accumulate along the way and in the ocean.

With the demands placed on the system by growing densities of people and homes, fairly raw septic seepage finds its way into aquifers, reservoirs, streams, and rivers. Every day, for example, more than forty million gallons of partially treated sewage pour into New York City's Hudson and East rivers. The public sewer system, though controllable at the waste treatment plant, also diffuses raw wastes from time to time in flood runoffs. Even after treatment, bacteria levels in wastewater often exceed acceptable levels.

Compared with the focus on pollution of rivers and lakes, which grew in force in the early 1970s, groundwater contamination is a fairly recent concern. Partly because we know least about it or how to clean it up, the pollution of our aquifers is a particular object of anxiety. Found in small interconnected spaces between soil and rock particles, both close to the surface and hundreds of feet under the ground, groundwater supplies drinking water to about half the country, around 40 percent of Americans. More than 90 percent of rural America depends on groundwater in aquifers for drinking water supplies, and also for 55 percent of the water consumed by livestock and 40 percent of that used for irrigation; aquifers also recharge surface waters, supplying about 30 percent of the water in rivers, so their contamination also affects the health of plants, birds, and wildlife.

For the most part aquifers replenish themselves so slowly that they constitute a nonrenewable resource: Some of the underground water supplies we are using now probably seeped into the earth as long ago as two hundred years. In some places, particularly on irrigated semiarid farmland, we are consuming our underground waters at a profligate rate. We are rapidly "mining" the aquifers of California's Central Valley and the vast Ogallala Aquifer, formed during the Ice Age, that waters 170,000 square miles of the Great Plains—particularly in the Texas Panhandle, where 25 percent of the Ogallala's preagricultural volume has been pumped away. An estimated 73 percent of water withdrawn from the earth is used for agriculture. Irrigation as practiced in many places is extremely inefficient: On average, only 37 percent of irrigation water is taken up by the crops; the rest can be considered lost.

People seldom become aware of groundwater pollution until some time after it has occurred. Drinking water can be contaminated by farm chemicals and still taste, smell, and look potable. Pollutants in groundwater usually remain concentrated in plumes or columns that spread slowly, and they may turn up in wells or reservoirs only years later. Because groundwater is cut off from much of the oxygen needed by microbes to break down wastes, its capacity for self-purification is very low. In addition, the ability of the earth to break down contaminants in water percolating through it is dramatically curtailed in the case of synthetic organic chemicals.

Cleaning up the underground pools once they are contaminated is expensive and sometimes impossible. Inserting gauges, tools, and hoses into an indeterminate underground crevice might instead destroy that space and spread the pollution. For the same reason, information about how far contamination has permeated is limited. What happens to the chemical wastes injected into the ground in more than 260,000 injection wells, for example, will remain mysterious until we have actual evidence of seepage. But we now know that water supplies under the earth are receiving the fallout from a vast array of land uses, including farm chemicals, industrial impoundments, hazardous and municipal waste landfills, underground storage tanks, septic tanks, oil well drilling, pipelines, and mining.

It is much easier to measure the pollution of surface waters and its impact on the health and survival of aquatic creatures in lakes, rivers, and seas, though not its impact on our own bodily well-being. Generally healthy in the 1950s, the Chesapeake Bay has been inundated since then by farm chemicals and manure as well as runoffs from heavy industry and housing development. In the bay where Captain John Smith was reported as saying in the early 1600s that there were so many fish he could catch them in a frying pan, the 1990 angling season was abruptly ended after a few days to protect the surviving fish. Worse yet, in 1992, the oyster catch was less than one tenth of what had been harvested at the turn of the century.

Up and down the East Coast of the United States, red tides were once extremely rare. Since 1972, however, toxic blooms of red algae, apparently nourished by heightened nutrient levels, have struck six times. Around the Great Lakes small but growing numbers of waterfowl now exhibit a pattern of birth defects including crossed beaks and club feet, apparently as a result of preceding generations of birds' eating fish afflicted with tumors and cancer of the liver. The effect of pollution on the survival of larger mammals was dramatically spotlighted by the *Exxon Valdez* spill, as the rate of reproduction of birds of prey and sea otters sharply declined, and many whales just disappeared.

Throughout North America one in three species of freshwater fish is threatened by degradation of lakes, rivers, and streams. Little progress has been made in reversing declines in fish species. Particularly in the arid Southwestern states, delicate aquatic habitats are being destroyed by sewage and industrial pollution, by diversion of water for drinking and irrigation, by farming, and by the introduction of competitive predatory species. Virtually all fish in the Southwest may be endangered because growing human populations are competing so strongly for scarce water. The Colorado River, which used to flow 1,450 miles from the Rocky Mountains to the Gulf of Mexico, now reaches the sea as a trickle, its waters snatched along the way by dams, farms, and thirsty people. Experts predict that extinctions will increase sharply unless public policy begins to deal with entire communities of fish and their ecosystems rather than focusing on individual species.

Louisiana, a state shaped by the Mississippi and rimmed with bayous, represents the worst example of the shortsighted bargain struck repeatedly in the postwar United States, ceding the "free goods" of air, land, and water in exchange for industrial development. Some sixty years after Governor Huey Long gave carte blanche to oil and petrochemical companies along the Mississippi, the seventy-five-mile stretch of river from Baton Rouge to New Orleans (home to seventy-eight chemical plants and refineries, including Exxon, Dow, Georgia Pacific, BASF-Wyandotte, Borg-Warner, ICI, and Allied Chemical) has been dubbed "cancer alley." More than five hundred abandoned hazardous waste sites add their effluents to current record-setting industrial emissions (120 million pounds higher than those of second-ranked Texas in 1988), draining into waters and seeping into underground wells.

The cumulative contamination that for many years made Louisiana the number-one industrial polluter among the fifty states has nearly destroyed the web of aquatic life in and around New

Orleans' mammoth Lake Pontchartrain. Fourth largest in the nation and once a particular source of Cajun pride, the lake has become a receptacle for every sort of runoff from accelerated residential and industrial development, including raw sewage, pesticides, and chemical residues. The oak trees that used to shade the banks are gone, and the marshy shoreline is eroding. Swimming has been prohibited for years, and where fishermen used to haul in a bounteous mix that included crabs, shrimps, alligators, and several sorts of fish, they now get mostly crabs and catfish. Down on the Gulf Coast, about fifty square miles of wetlands disappear each year, their vegetation blighted by canals dug for oil exploration.

Coastal wetlands are particularly vulnerable to the inroads of development. Our society's discovery of the value of these natural areas (defined as areas saturated by surface or groundwater with vegetation adapted for life under those soil conditions, as in swamps, bogs, fens, estuaries, marshes) came belatedly, after we had lost about half of the two hundred million acres in existence when the first Europeans arrived. Our new appreciation of wetlands is signaled by the use of the word itself, replacing "swamp" or "marsh." (The word "wetlands" is not to be found in H. L. Mencken's *American Language* of 1948, although more than thirty meanings of the word "swamp" were included.) Formerly considered soggy wastelands that could be made useful only through the ingenuity of drainage engineering, coastal marshes are now known to fulfill a number of beneficial—indeed indispensable— natural functions. They provide an essential habitat for countless species of mammals, birds, fish, and invertebrates; feed and shelter at least 90 percent of fish and other marine species at some stage of their lives (coastal mangrove swamps and salt marshes are the earth's zone of highest biological productivity); process and filter wastes; slow and store floodwaters; and protect shorelines from erosion. Wetlands generally constitute a buffer between urban and agricultural runoff and surface waters, crucial in maintaining water quality.

As pollution and development consume some 350 to 650 square miles of wetlands a year, Americans are engaged in a tug-of-war over the shoreline. The debate reflects the fact that the value of water-edge property as real estate is still easier for many people to grasp than the high economic stake we hold in common in nurturing species and their habitats. In Florida, however, the value of wetlands and the difficulties of reclaiming what is lost in tampering with them has been painfully brought home by the partial destruction of the Everglades, southern Florida's four-thousand-square-mile riverine wetland.

Whatever the relative responsibility of farmers, developers, and the Army Corps of Engineers flood control project, as a result of their combined activities, by the late 1980s the Everglades had shrunk to half its original extent and had lost to pollution and predatory competitors much of its teeming population of coots, ducks, wading birds, alligators, fish, trees, and plants. In the sixties the Corps redesigned the Kissimmee River, transforming its meandering, grassy waters into a straight, deep canal, promoting expanded farming and residential construction; by 1971, heavily polluted farm runoff plunged straight down that channel into Lake Okeechobee, fouling the state's prime source of fresh water and choking life in the Everglades beyond. In 1992 the Corps decided to try undoing its handiwork and set the river free again within its original course. Even if politics permitted such a radical reversal, however, the ability of the enervated, contaminated, and colonized Everglades marshlands to spring back remained in considerable doubt.

Looking beyond our coastlines at the final repository of all these runoffs, we can take some solace in the robustness of the great oceans, the first source of life, whose extent and power dwarf all else on this planet. Covering 71 percent of the earth's surface, the oceans play an essential role in maintaining the conditions for existence on land, moderating temperature and moisture by absorbing heat and carbon dioxide and giving fresh and pure water back to the atmosphere through evaporation. They also feed, shelter, and nurture a vast world of living organisms, which account for a goodly share of our food supply. Despite all the residues of modern human life that pour into the open oceans, their enormous diluting power appears so far to have protected their physical and chemical environment. As we have seen, however, the coastal areas and regional seas and the marine habitats within them that nurture ocean life have not fared so well.

Basic elements in coastal ecosystems have been altered, and some impaired irreparably. Stands of mangroves have been lost, and delicate coral reefs, home to an estimated one million species (including two thousand varieties of fish), are swiftly falling prey to the toxic side effects of development and climatic change. In Costa Rica, 75 percent of the reefs have been killed by sedimentary runoffs from local rivers; in the Philippines a hundred thousand jobs are lost each year as fish disappear from reefs, killed by pesticides, erosion, and cyanide used to stun exotic specimens for aquariums; in Indonesia, Kenya, and elsewhere that cycle will inevitably follow as fishermen dynamite reefs to flush out their catch. Around Sri Lanka, divers strip-mine the reefs to use in making plaster and cement. Marine life in coastal waters commonly suffers from a lack of dissolved oxygen caused by an overload of waste nutrients. Along the East Coast of the United States, low oxygen has led to sharply declining catches of fish and shellfish; during the summer months, the heat and nutrient growth combine to produce "dead spots" virtually devoid of oxygen and marine life. Vacationers have become accustomed to bans on local shellfish even as they seek the beauties of seaside air and water.

Less evident but equally serious is damage to the sea's extraordinarily rich surface microlayer. Marine biologist John T. Hardy underlines the key role of "this complex skin of the seas" wherein "dwell thousands of plants, animals, and microbes, all attracted there by its special ability to nurture life." The layer is densely packed with microalgae and protozoa and the larvae that feed on them, as well as the buoyant, fatty fish eggs of myriad species. The surface waters also contain a complex, concentrated mixture of chemicals, such as amino acids, proteins, and fatty acids excreted by the tiny plant and animal organisms, which forms a kind of film. Sulfide gases exuded by plankton rising from the surface also influence weather, by serving as nuclei for cloud condensation.

Unfortunately, contaminants dumped in the water or falling from the atmosphere also collect in the natural organic film as nonsoluble pollutants bind to buoyant particles. According to Hardy, studies in Puget Sound, Chesapeake Bay, the North Sea, and the waters of Southern California and Florida show that the microlayer "is becoming a soup of toxic metals, organic pollutants, bacteria, pesticide residues and the byproducts of combustion-derived hydrocarbons from cars, trucks, airplanes, refuse incinerators" and sewage plants. Not surprisingly, fish eggs developing amid this pollution stand a good chance of dying or emerging malformed. Because the dense congregations of microbiota and larvae feed both seabirds and other creatures swooping from above and many larger organisms swimming upward from the deep sea below, the poisons in the surface layer migrate throughout the marine food chain. During the past decade fish and marine mammals have been afflicted with an unprecedented rash of infections, ulcers, cancers, and epidemic viruses.

Both pollution and the pressure of more fishermen sweeping with ever-larger nets to supply food to more people are taking a noticeable toll on annual catches. Open-ocean drift nets up to forty miles long—large enough to catch Manhattan—have dragged the seas for more than a decade. Although most maritime nations agreed to a U.N. resolution banning the nets in 1993, overfishing continued to threaten sustainable yields in coastal waters. The per capita global catch trebled between 1950 and 1980 but has not grown greatly since. In 1992, it was down to 97 million tons from its 1989 high of 100 million tons. Scientists are now asking whether the yearly harvests of the fifties, sixties, seventies, and eighties were not exceeding sustainable yields.

American fishermen from Booth Bay Harbor, Maine, to Sea Island, Georgia; from Seattle, Washington, to San Diego, California, are feeling the pinch. In Gloucester, Massachusetts, in 1992, an expanded fleet had to stay out twice as long to bring back half as much fish from Georges Bank, as stocks of cod, haddock, and flounder sank to their lowest level ever. With fishermen bringing in an unsustainable 50 percent to 70 percent of available stocks, they were being forced to submit to regulations limiting their days at sea, decreasing their net size, and banning newcomers to the fishery. For the foreseeable future, the sometimes perilous but always freewheeling days of coastal fishing are over.

The costs of the total damage our cumulative wastes have inflicted are daunting. Quite apart from the monies we may have to spend for cleanup, we have already irrevocably lost some aesthetically and economically sustaining aspects of our world. We will continue to lose fish and plant species, forests, lakes, wetlands, aquifers, and available lands until we adopt a conservationist approach to economic progress. As a society we are already beginning to accept the reality of the environmental costs. What is likely to be more controversial from now on is the psychic cost of radically changing our economic lives to curtail waste and pollution.

Part II

Managing
Garbage

6

~~~~~~~~~~~~

Nowhere to Hide It

Every advantage in the past is judged in the light of the final issue.

—DEMOSTHENES, *First Olynthiac*

A S WE BECOME more aware of what we have wrought, people are beginning to understand how far the limits of our material lives are defined by the problem of waste. Ultimately, we will only be able to deal adequately with the fallout from both consumption and production by shifting our focus from waste management to reduction. Getting from where we are now to a full-fledged effort at waste reduction clearly will represent a real leap. Because for most of us garbage represents our most direct involvement with the environment, how we handle our household trash will help shape our response to the larger issues. Before grappling with the fundamental question of how to reduce waste on a societywide basis, therefore, we will look first at how we are doing on the household garbage problem.

Since the municipal housekeeping reforms instituted in the 1890s, technologies and practices in waste disposal have shifted back and forth among the three time-honored alternatives: dumping, burning, and reusing. High-volume waste problems have also shifted radically, from the manure glut of the nineteenth and early twentieth centuries to the ash deluge of the wood and coal eras—in 1939 ashes made up 43 percent of household waste—to the paper storm, packaging, yard waste, and toxic residues we struggle with today. For our first several hundred years, pigs were the principal garbage collectors and recyclers, both in cities and on farms. Toward the end of the nineteenth century and until World War II, many cities also opted for incineration plants as the way of the future. Because of our vast landscape, however, dumping has always been a relatively easy option for Americans.

In the postwar era, while Japan and many European nations invested substantially in incineration, the United States relied on landfills to dispose of most municipal solid waste.* Thus, when our contemporary garbage problem made itself felt in the mid-1980s it was at least initially a landfill problem. New York City's Fresh Kills landfill, soon to reach capacity, represents an extreme example of the trash problems that have piled up during our boom years.

In other times, the Fresh Kills dump might have ranked as one of the world's wonders—a hole in the ground that became a mountain. Sere and sandy, its towering bulk stands like a desert peak risen from the marshes of Staten Island. Thirty-four million pounds of refuse arrive daily at Fresh Kills via a fleet of barges, while a land army bulldozes the trash skyward, dumps dirt over it, and packs down roads in the path of trucks and equipment. When it is completed around 2005, it will stand five hundred feet tall and cover more than a thousand acres.

Despite Fresh Kills' claim to a number of superlatives, in many ways, the history of this landfill typifies the evolution of our national municipal waste dilemmas. After World War II, reaction against belching incinerators, which by then consumed the trash of some seven hundred U.S. cities, coupled with the increasing

---

*In 1987, incineration accounted for 50 percent of solid-waste disposal in Japan, 30 percent in Germany, 36 percent in France, 40 percent in the Netherlands, 55 percent in Sweden, and 70 percent to 80 percent in Switzerland (compared with around 7 percent in the United States). Correspondingly, landfilling accounted for only 16 percent of solid-waste disposal in Japan, 30 percent in Germany, 36 percent in France, 40 percent in the Netherlands, and 5 percent or 10 percent in Switzerland (while the United States continued to landfill about 82 percent).

availability of truck transport, led municipal sanitation departments to the cheap real estate at the edges of town. In the eu-
phemistic spirit of the early 1950s, they coined the phrase
"sanitary landfill" to describe the modern dump—where garbage would be neatly hidden by a regular layering of dirt. In fact,
many landfills were more like open pits, infested with vermin and
smoky with flash fires.

In 1949, when Fresh Kills first began receiving New York City's
trash, city officials described the new dump as a sort of urban
renewal project wherein the "unsightly and unsanitary wastelands" in Staten Island could be filled in with garbage and then
transformed into a sweep of parks. The gap between that bucolic
vision and the malodorous leaking mountain of today is striking,
not only in terms of the pile of waste swollen beyond the wildest
imaginings of the original proponents, but because of our in-
ability to deal with it adequately. This disjunction gives a measure
of the changes during the last four or five decades and of how
we view what has occurred.

The Fresh Kills landfill aptly represents what our fifty-year
binge has wrought. Every day more than a million gallons of
liquid, spiked with a chemical mix including household toxics,
commercial solvents, and automotive fluids, leach through the
trash mountain into Arthur Kill Bay and the network of underground waterways beneath the fill. Because the dump lacks a gas
collection system, methane and other gases exuded by the decaying garbage, some of which compound the greenhouse effect,
escape into the air. Although the problems of New York City's
landfill are outsize, in its lack of environmental safeguards it resembles most existing municipal dumps.

As concern for pollution has intensified, government responsibility for waste has been more widely shared. Local government
has always had responsibility for dealing with solid waste disposal,
but a significant federal coordinating role originated in the
1970s. Public and governmental commitment on national and
local levels has fluctuated, as Congress and the state legislatures
have shifted their focus from one area to another, passing laws
to curb pollution of the air, water, land, and then air again.

The Resource Conservation and Recovery Act of 1976 (RCRA)
definitively changed the world of municipal solid and hazardous
waste. In the case of garbage, the law promulgated the first fed-

eral standards for waste disposal in landfills by states and municipalities.* While outcomes under RCRA have fallen short in many ways, the assault it launched on pollution has changed the way we act and think.

The perception that pollution is waste and that waste pollutes has formed the basis for declaratory policy on waste disposal, in RCRA as in the earlier Resource Recovery Act of 1970. Throughout the 1970s and '80s, federal and state legislation repeatedly affirmed a hierarchy of waste management alternatives, starting with resource recovery and recycling, and descending to incineration and, finally, to landfilling. The priority put on resource recovery sounded loud and clear in the titles of both laws passed by Congress in the 1970s.

For the most part, the emphasis on reduction has been entirely rhetorical. During the two decades after 1970, the real priorities for government action remained essentially the reverse of the rhetorical ideal: Landfills dominated the terrain, followed by incineration, followed by recycling. For waste managers, both landfilling and incineration represent ways of making garbage disappear without looking at it too closely or changing the way it is collected. Landfilling and incineration seem appealingly simple compared with recycling and source reduction. Thus, for some years the preferred alternatives were almost completely ignored.

Among environmentalists, industry, and officialdom, fashions in terminology shift. In surveying the alternatives, we will use some shorthand to bypass occasionally complicated semantics. When the EPA reaffirmed the place of the conservationist alternative at the top of the hierarchy in the latter 1980s, it used the phrase "source reduction" instead of "resource recovery," which the waste disposal industry had appropriated as a euphe-

---

*In the decade following the law's passage, RCRA's general prohibition on "open dumps" and the requirement that landfills should be "sanitary" (that is, disposal of waste at such a facility should not cause "adverse effects on health or the environment") prodded the states into closing nearly two thirds of existing landfills.

mism for incineration.* Source reduction (also called, in reference to industrial toxic waste, "pollution prevention") aims at preventing the generation of waste and thus differs from recycling, which manages waste after it has been generated. Like recycling, however, source reduction aims to reduce waste through more efficient use of resources—specifically by changing consumption patterns and manufacturing processes. For our purposes, "waste reduction" is an umbrella term for all conservationist alternatives including recycling, "source reduction," and reuse.

Since the mid-1980s, citizens acting at the local level have signaled strong support for the hierarchy as stated (with recycling standing in for most other forms of waste reduction). The result is the small but unmistakable shift we have seen, away from landfilling and toward recycling and composting, with a lesser growth in incineration as well. Estimates of the relative shares vary, but trends during the last decade clearly reflect sizable shifts in public attitudes and real efforts to realign habits, policies, and institutions. In 1980, according to the main EPA source, Franklin Associates of Prairie Village, Kansas, 81.4 percent of all municipal solid waste was landfilled, but by the end of 1990 that share had dropped to 66.6 percent while recycling had increased from 9 percent in 1980 to 14.9 percent in 1990. At the same time incineration increased from 9 percent in 1980 to 16 percent in 1990. More resent estimates derived on a somewhat different basis through a survey by the journal *Biocycle*, put landfilling's share at 72 percent, incineration at 11 percent, and recycling at 17 percent in 1992.

The trend toward recycling is still fragile. Programs vary greatly from region to region and state to state, with the strongest responses found, not surprisingly in regions where landfill dumping fees (known as "tipping" fees) have risen the most. In 1991 aggregate totals for recycling collection have risen to 22 percent of total waste disposal in New England, 18 percent on the West Coast, and 15 percent in the Mid-Atlantic area, while the Rocky

---

*The industry connotation also prevailed at the EPA in the late 1970s and early 1980s; the EPA did, however, unequivocally endorse source reduction at the head of the hierarchy in 1989, followed by recycling, incineration, and landfilling.

Mountain states and the Midwest continued marginal. State programs ranged from 34 percent of totals in Washington State to just 2 percent in Tennessee. But recent surges in numbers of people reportedly served by curbside recycling, from forty million in 1990 to more than seventy-eight million at the end of 1992,* and the doubling in the number of collection and sorting facilities in the same period, indicate real nationwide momentum.

To sort out the welter of data on materials, energy, and pollution in regard to all three alternatives, an analytical tool called "life-cycle analysis" (or "assessment," known in the trade as LCA) has achieved new prominence. As a template to help rationalize choices, LCA joins "cost-benefit analysis," which is aimed at comparing the costs and the benefits of pollution reduction, and "risk analysis," which aims at measuring risks to human health in quantitative terms, and sometimes at measuring environmental risks more broadly. Unlike cost-benefit analysis, which views the environment as a subsystem of the economy, and risk analysis, which considers pollution as one of many threats to the human body, life-cycle analysis takes the whole ecosystem as its field of concern and measures the "cradle-to-grave" impact of products and processes on that system.

The days of cheap and easy dumping and burning are clearly over. Land for fills has grown scarce as suburbs have tacked on their own suburbs and fears of pollution have grown. As a result of rising land costs and the upgrading required by environmental regulations, average tipping fees paid by haulers for dumping in landfills doubled between 1982 and 1987, from about $10 to $20. Between 1986 and 1990, increases ranged from 43 percent in the South to 269 percent in the Northeast. Nor had incineration offered a way to avoid rising costs, as tipping fees at incinerators rose from around $13 to $33.50 between 1982 and 1987. Some places felt the crunch keenly: In Philadelphia, for example, the

---

*Because fifteen states offering curbside recycling do not report population figures, the real totals are probably closer to 85 or 90 million.

cost of garbage disposal grew to exceed expenditures for fire protection, rising from $8.5 million in 1983 to $40 million in 1988.

Rumors of the demise of landfilling must, however, be viewed as premature. Although hopes have risen that recycling will offer an economical alternative, dumping is still the lowest-cost form of waste disposal for most municipalities.* It is true that when Fresh Kills puts a final cap on its summit in a few years, around 80 percent of the country's *currently* operating municipal landfills will indeed have closed; for a few years, at the end of the 1980s, overall landfill capacity seemed to be sloping swiftly downward— during the 1970s and 1980s, for example, almost 70 percent of all landfills (fourteen thousand facilities) closed. Yet, as William Rathje and Cullen Murphy pointed out in *Rubbish!*, landfills are always closing. As they see it, at any given moment perhaps half of all existing landfills will close within five years. Only in specific localities do we actually seem to be "running out of space." In fact, as a result of increased competition within the garbage industry and a spate of activity by municipal officials, capacity in large, new, state-of-the-art landfills may exceed what was lost with the closing of all the small dumps. America's towns and cities are currently consigning trash to approximately six thousand dumps nationwide.

During the 1980s state laws began the process of corseting and grooming the blowsy dump of old. In September 1991, EPA standards mandated by Congress in 1984 made the modern era official nationwide. All new landfills were required to line the pits, typically with plastic sheeting on top of layers of clay, to prevent leakage; to collect liquids that do leak; and to monitor the production of methane gas produced through decomposition of leaves, grass, food, and other organic wastes. In addition all were enjoined to test groundwater for leakage of lead, plastics, or other chemicals, as only about a quarter do currently. Owners were also now required to monitor dump sites for thirty years after they ceased operations.

---

*Figures quoted here do not generally include construction and demolition debris, which may make up a large portion of wastes going to landfills (perhaps 25 percent to 30 percent of the municipal waste stream, or thirty million tons, according to garbage analysts, Franklin Associates of Prairie Village, Kansas.

As the new regulations take hold, they should mitigate the risks of pollution from landfills built after the legislation. The RCRA regulations support state legislation during the past decade that will presumably substantially curb contamination in new landfills. Yet the effect of the regulations will be more limited than their authors hoped. The law allows states considerable leeway in carrying out the rules after the EPA approves their programs for permitting landfills. Although about half the states already had analogous landfill regulations before the new federal standards were promulgated, enforcement at the state level was haphazard. Many states simply lacked the personnel and resources to maintain effective inspection.

In addition, the durability of various sorts of liners, constructed from synthetic and clay materials, cannot be predicted with certainty. Because all will eventually deteriorate,* and gases can be produced in the landfill for decades after closure, as long as dumps are part of the waste equation, the threats to groundwater and air cannot be effectively alleviated by state-of-the-art liners, pumps, pipes, or vents alone. In the long run, the only way to cut down contamination in landfills is to change the composition of the wastes we deposit there. Contamination of landfill leachate or compost could be greatly reduced, for example, if the dumping of household hazardous wastes such as paints, solvents, oil, cleansers, and batteries†
were eliminated.‡

---

*Estimates range from ten to fifty years, depending on who's estimating, with an average of around thirty. Thirty years is also the length of time that landfills are required to continue monitoring after closure. No science, just good round numbers.

†Europe is way ahead of the United States in battery collection programs and laws mandating reduced levels of heavy metals. During the 1980s, European standards drove American battery manufacturers to cut back mercury in alkaline cells by over 90 percent, and by 1993 EC standards will cut both mercury and cadmium in alkaline batteries to .025 percent. (Button cells used in hearing aids, watches, and calculators and the nickel-cadmium batteries in small rechargeable appliances will remain significant heavy-metal sources.)

Japan and a number of European countries collect batteries and occasionally other household toxics for recycling and disposal as hazardous waste. Both Japan and many European countries also collect electrical equipment and "white goods" like refrigerators and washing machines separately.

‡Improper dumping of industrial waste has also played a role in the severe pollution of municipal solid-waste landfills. The disposal of incinerator ash in regular landfills poses another potential pollution problem.

☙

If incinerators worked more efficiently and didn't cost so much, they might be expected to rank as odds-on favorites in a popular choice among alternatives. Incineration is the favored technology of many municipal officials, and of the waste disposal industry. These big, new, state-of-the-art machines that can turn refuse into electricity or fuel pellets and bring economies of scale to the trash problem ought to represent modernity and efficiency. More than landfilling or recycling, incineration claims to *get rid of the garbage*—make it disappear. All one has to do is haul the stuff in and burn it. You don't have to go through a big fuss about changing institutions and lifestyles.

During the first half of this century, incineration seemed to provide the easiest answer to the garbage problems of rapidly growing cities. Then as more recently, the technologies were not invented here, but imported from Europe, where disposal systems developed earlier, from the late 1880s onward. Much money was wasted while we tried to burn American garbage in incinerators designed for drier European refuse (of the 180 incinerators built in the United States between 1885 and 1908, 102 had ceased to operate by 1909). After 1910 and through the 1930s, however, between six hundred and seven hundred cities had built incineration plants.

Between World War II and the present, incineration's prospects have risen and fallen several times. After its steep postwar decline, incineration appeared to be rising again, phoenixlike, in the 1970s. This time the guiding concept was "waste to energy." As before, however, we got into trouble in trying to adapt European technologies to serve very different American needs, for example, by increasing the size of the plants to achieve economies of scale and up the output of energy. A greater proportion of metals and particularly plastics in American than in European garbage led to corrosion of the mechanisms and increased air pollution. Most new incinerators were beset by cost overruns and mechanical breakdowns. By the end of the 1970s incineration had again declined and was handling less than 1 percent of all U.S. municipal solid waste.

Even then, however, tightening restrictions on landfills and

the arrival of the garbage crisis were again reviving the incineration industry. The Carter administration's alternative energy
program gave incineration a boost with the passage of the Public
Utilities Regulatory Policies Act (PURPA) of 1978, which mandated the purchase of energy by utilities from independent generators at a price equivalent to the most expensive source. From
the late 1970s onward, the Department of Energy and the EPA
pushed for increasing "waste-to-energy" capacity, and the DOE
called for the construction of 200 to 250 new incineration plants,
to consume 75 percent of the nation's municipal solid waste by
1992. The industry also benefited greatly after 1980 from the
Reagan administration's drive for privatization of municipal services. The tax incentives, loan guarantees, below-market loans,
price supports, energy entitlement grants, and industrial development bonds that became available through that initiative constituted a significant subsidy for incineration.

Political scientists Louis Brumberg and Robert Gottlieb have
chronicled the industry's abrupt shifts of fortune. By the mid-
1980s, once again it appeared that incineration's prospects knew
no bounds. The first half of the decade saw a burst of orders for
waste-to-energy plants, many of which came on line in the latter
part of the 1980s and early 1990s. Within five years the number
of plants already built or under construction had tripled to nearly
two hundred. In a 1986 study, Kidder, Peabody cited the large
incinerator plants as one of the few growth areas for engineering
and construction firms, and in 1987 *Forbes* magazine valued the
waste-to-energy market at $35 billion. Not surprisingly, the enor-
mous amount of municipal financing invested attracted a broad
coalition of proponents, including investment bankers, bond
brokers, and municipal organizations like the U.S. Conference
of Mayors, in addition to vendors, waste haulers, and component
manufacturers.

By 1991, the nation was back to burning around 14 percent
of its municipal solid waste. Yet the surge of investment in incineration had tapered off. Many of the tax advantages had disappeared with the enactment of the Tax Reform Act of 1986. With
the explosive growth of opportunity, many of the companies serving as project managers of the plants were new entrants into the
incineration business and had little operational experience. The
four largest players in the construction and operation of nuclear

plants undertook or increased their roles in the waste-to-energy field during this period. As Brumberg and Gottlieb and others have pointed out, there are many parallels between the nuclear power and waste-to-energy fields. Both require heavy capital investment, involve large-scale and to some extent unproven technologies, have enjoyed substantial support from the federal government, and have over time encountered major cost and environmental problems.

As in the 1970s, public patience wore thin as inexperienced managers battled the endemic mechanical problems of the complex new technologies. By 1988 mechanical failures had caused shutdowns at more than half of the operating plants and forced numerous closures. Doubts were rising about technology, costs, and environmental impacts. At length, municipal sanitation officials found themselves pulling in one direction while citizens groups pulled in the other. Between 1987 and 1991, plans for 121 proposed waste-to-energy facilities were canceled.

The saga of Los Angeles' abortive ten-year quest for mass-burn incineration was emblematic. In the late 1970s, after the residents of the Santa Monica Mountains area protested landfill expansion in neighboring canyons, the L.A. Bureau of Sanitation moved ahead with little public scrutiny, laying plans for three incinerators (under the rubric Los Angeles City Energy Recovery Project, or LANCER). In December 1984, prior to any public review process, they sited the first of the three plants in a deteriorating minority neighborhood in the south-central area. In the spring of 1986, with the whole package in place, they held community workshops to "simulate opposition," in case any later court action raised questions about the hearing process.

At the start of the hearings, neighborhood residents were confused; according to a *Los Angeles Times* report, "some of the elderly people" thought LANCER was "a shopping center." A year later, however, the project that Los Angeles had heralded as the "21st Century Solid Waste Management Solution" and invested $12 million to develop was dead in the water. Problems and uncertainties dragged out from under the rug included additional expenditures for infrastructure, water, sewage, shortfalls in energy revenues, and additional pollution controls, which would raise costs per ton from $27.43 to $56.89; doubt about assumptions of health studies regarding dioxin levels, acidic gases, and existing air pollution; and what the public felt were glaring omissions in the city's environmental and cost studies (principally regarding ash residues). At the end of 1990, city officials reportedly were planning once again to landfill at least half the city's garbage. In 1994 they were briskly filling in the adjoining canyons—while also moving ahead on recycling.

Analogous events have been repeated on a smaller scale in various cities around the country. In some cases, such as the impasse over a proposed ashfill in Rutland, Vermont, communities have decided not to pour any more money into what is now seen as a losing venture. In others, as in Jackson County, Michigan, authorities have closed plants found to exceed pollution levels in air or ash. In 1987, several newly elected city councilors in Austin, Texas, canceled an incinerator under construction, on the rationale that it made sense to write off the $23 million already spent because over twenty years the city would save more than $100 million by investing in recycling. More frequently, communities have simply absorbed the unforeseen costs, as did Warren County, New Jersey, for the disposal of ash with unexpectedly high levels of toxicity; or, most typically, as did Hartford, Connecticut, for mechanical problems, including burst steam pipes, shortly after the startup of its new incinerator.

As American technicians gain more experience, incineration's problems may be tempered. Incinerators *do* reduce the volume of trash that must be landfilled—given the characteristics of the U.S. waste stream, anywhere from 90 percent to 60 percent. In time, mechanical failures and pollution could be mitigated in part by significant progress in separating out metals and other toxic materials through pollution prevention and recycling. In addition to the elimination of batteries and metal objects, removal of plastics, glass, and pigments in paint and ink would greatly reduce the level of nitrous oxide and heavy metals; elimination of corrugated boxes, rubber, and plastic would curb sulfur dioxide; and the weeding out of certain kinds of plastics would reduce chlorinated organics.*

---

*Paper itself is a significant source of both nitrogen oxide and sulfur dioxide.

Nonetheless, incineration's environmental problems will never be "solved." That is because the plants will continue to chemically transform and concentrate the toxics in the waste stream. The result inevitably is toxic ash to be landfilled or particles emitted through smokestacks.

Proponents of incineration often assert that current emissions and combustion control technologies can eliminate all but minuscule fractions of air pollutants, but data supporting this claim is scanty. Although we have reason to believe—and some evidence, as well—that recently built facilities have significantly curbed both air and ash pollution, existing information gives us very little basis for certainty in drawing conclusions. Industry measurements and projections of incinerator air pollution almost undoubtedly underestimate the risks. Government monitoring is spotty at best, and most risk assessments are based on sparse data derived from a few facilities under optimum conditions—and very often supplied by the proprietor. Thus they often fail to account for conditions commonly attendant upon incineration, including tremendous inherent variability in the composition of municipal solid wastes, frequent temperature changes, instabilities during startup and shutdown, and aging of machinery. Nor do they take into consideration the training and professional commitment of personnel, which also affect operations profoundly: Fluctuating attention and skill of operators can cause wide variations in combustion and in outcome.

Many incinerators, even with the latest equipment, still emit significant amounts of criteria pollutants.* Recently, a major study by the environmental research organization INFORM comparing "state-of-the-art" and actual practice showed that although newer plants perform better than those built a few years earlier, few in fact reliably meet emissions levels that are demonstrably achievable using currently available technology. Considered individually, the health effects of given levels of each criteria pollutant may be debated. In addition, the most controversial gas emitted in incineration, dioxin, is not defined as a criteria pollutant. Dioxin is alternatively characterized as the most potent

---

*Six pollutants known to be hazardous to human health (ozone, carbon monoxide, total suspended particulates, sulfur dioxide, lead, and nitrogen oxide), used as guidelines for setting standards for allowable concentration levels.

carcinogen ever tested in laboratory rats and as a minimal risk
to humans in small amounts. But little is known about the cumulative effect of any of these pollutants over time, particularly
within the crowded urban cachement areas where incinerators
are often located.

Interim air pollution standards for large municipal incinerators were issued for review in 1991 and are to become effective
after that review in 1994; the standards should help, but will still
leave many problems unsolved. They will for the first time bring
some agreement on what emissions are measured, how they are
measured, and what set of operating conditions should be defined as a standard at least for criteria pollutants. Nonetheless,
confining measurement to only the targeted criteria pollutants
will leave many pollutants unmonitored. The new standards also
fall short of what can be achieved with current state-of-the-art
technologies.

The failure to promulgate standards for toxics in ash at the
federal level and in most states does not reflect the relative risks
of ash versus air pollution, but, rather, strenuous lobbying against
characterizing ash as hazardous waste. Although incinerator ash
is not now included in tabulations of municipal solid waste, it is
commonly dumped in municipal landfills with the rest of the
garbage. And both municipal officials and the incineration industry want to keep it that way. The gist of current arguments
for the treatment of ash as regular municipal waste emphasizes
that it is "less hazardous than hazardous." What comes through
much louder and more clearly from those making this case, however, is the question of the bottom line: Segregating ash makes
it much more expensive to dispose of. Studies of ash under various conditions have shown it to be hazardous by numerous
measures. Federal, state, and local government and industry studies of ash leachability, according to EPA's extraction procedure
toxicity test, show that virtually all fly ash (noncombustible resid-
ual particles expelled by flue gas in stacks) tested exceeded federal standards, usually for both lead and cadmium, and at nearly
half the facilities tested, mixed bottom (nonairborne combustion
residue and noncombustible materials) and fly ash exceeded federal standards for lead.

Looking at incineration's role in competition with recycling,
the fears of recycling proponents that incinerators hungry for

trash and for fees paid by municipalities will slow down waste prevention and recycling programs appear well founded. In 1990, the New York town of Hempstead, Long Island, found that if it achieved its goal of recycling 25 percent of its trash it would not meet its contractual obligation to send 540,000 tons a year to the locally based American Ref-Fuel incinerator. Thus the town faced financial penalties for cutting its waste. In Babylon, New York, in 1989 half the newspapers collected for recycling were burned instead so the town could supply half the tonnage it had guaranteed to the local incinerator. And the Claremont, New Hampshire, incinerator incurred a $64,000 deficit in the first quarter of 1990 when it fell short of its refuse quota as a result of recycling efforts in the twenty-seven towns feeding into it.

In New York City, a long-running battle over waste management strategy has pitted the City Council, which has favored recycling, against the Sanitation Department, which traditionally supported "integrated waste management," that is, incineration. In the summer of 1992, as sanitation workers mopped up trash-strewn streets before the Democratic Convention in early July, the City Council rejected a proposed three-thousand-ton incinerator in the Brooklyn Navy Yard for the umpteenth time. In a city with almost no empty land except in Central Park and abandoned areas of the South Bronx, and boasting a trash output of 26,000 tons a day and an overflowing landfill, such a decision might seem foolhardy. The City Council's action, however, made sense in the context of the guerrilla politics that has come to dominate the debate about incineration.

In early 1994, the deal finally struck a year and a half earlier offered some hope that waste reduction would begin to play a role in New York: While the Brooklyn incinerator project was going ahead, after several years of halfhearted efforts to establish curbside recycling, a citywide program was introduced in the fall of 1993. However, the Sanitation Department's view of the power hierarchy seemed manifest in its allocation of resources: In fiscal 1994, only 10 percent of the $641 million Sanitation Department budget was allocated to recycling. The tussle over New York's incinerator is about forcing city officials to put their money where their rhetoric is—and commit to recycling.

Because waste reduction not only conserves materials and energy but also bypasses waste disposal, it offers the most cost-effective way to handle waste and pollution. Both dumping and burning have been indispensable in maintaining the habitability of urban living spaces throughout human history. Both still have numerous proponents, who tend to treat recycling as well-meaning but "unrealistic." Landfill advocates say that whereas incinerators will inevitably either pollute the air or intensify toxics in the ash, or probably both, landfills with liners are almost harmless. Proponents of incineration assert that while all that stuff molders in eventually leaky landfills, high-tech, clean-burning incinerators simultaneously reduce waste and recapture energy.

In fact, both dumping and burning make extremely inefficient use of land, air, and materials. As we crowd our biosphere, we can no longer afford this prodigal use of resources. Because we are unlikely ever to achieve perfect efficiency, our need for the two disposal alternatives will never end entirely. When circumstances force us to resort to either method, however, we must work constantly on improving environmental safeguards. As we pursue both economic and environmental progress, phasing out both landfilling and incineration should be our goal.

7

~~~~~~~~~

Choosing Conservation

Less is more.

—Ludwig Mies van der Rohe

W ITHOUT EXACTLY DEFINING or acknowledging the attraction, many Americans who are worried about waste and fearful about overloading the system are drawn anew to the idea of thrifty husbandry of resources. Their response is characteristic of the traditional national zeal for individual problem-solving. Recycling has become popular because it gives us a way to *do something* about waste on a personal level.

This new conservationism has very fragile economic roots, however. Without a much more radical change in course, efforts to relieve stress on our environment through recycling and other forms of waste reduction may well go the way of other innovative programs ultimately dismissed as "unrealistic." Only if we can more fundamentally change the rules of the game that govern our use of resources are we likely to extend current good works into a broader program to stem the crisis.

For Americans, the word "conservation" itself still constitutes something of an obstacle. For many of us, the term still connotes a reining in of the expansiveness that has characterized our quest

157

for wealth and conquest of this continent. So proponents of environmentalism are often wary of invoking the concept. Though at present the U.S. public decidedly prefers recycling to other waste management alternatives, for sensible political reasons, knowledgeable advocates of the process don't use the word "conservation" much, avoiding possible associations with elitism, government interventionism, and hostility toward growth. Without defining it, however, many advocates of recycling are probably attracted to something broader—more than merely grinding up old bottles or melting used cans, they are saving materials. Thus the appeal of what is called "recycling" implicitly includes the other methods of avoiding waste and pollution—source reduction and reuse.

Really reducing waste, as we will see, will require far-reaching shifts in our underlying economic policies. On this spectrum, recycling is significantly less threatening to the status quo than source reduction and reuse, for where recycling works on waste after it has been produced, and is itself a form of production—fabricating objects while using materials again—source reduction changes the ways we produce and consume to prevent waste. For that reason source reduction and reuse are ultimately more efficient than recycling. Currently, however, recycling is our most widely understood approach to waste reduction, and for most people the only way they know of reducing waste and conserving resources. Thus it is worth exploring in some detail how it's working now.

Although the current level of popular support does not by itself ensure the early integration of conservationist alternatives into our economic institutions, large numbers of Americans clearly now want to make at least recycling work. In various polls, sizable majorities say they would prefer to purchase goods in recycled containers, and as a result of public pressure most states have set recycling goals. Over the past few years, while landfilling rates have dropped, and incineration remained stable, recycling has grown steadily—from 9 percent in 1989 to 17 percent in 1992. By that year, thirteen states had adopted laws requiring recycled content in various commodities.

Recycling seems so un-American that its popularity demands an explanation. A few years ago, it appeared unlikely that the idea of collecting materials for reuse would catch on here at all:

Europeans and Japanese, with long-ingrained experience of re-source scarcity, might be expected to sort their garbage, but this intrusiveness and constraint seemed likely to infuriate born-free Americans. We generate more waste than any other people in the world, and, as far as we know,* always have. Furthermore, up until recently, we have been rather proud of it. Yet, until about fifty years ago, recycling and source reduction were integral to most Americans' economic lives.

For the first several centuries of European settlement on this continent, economic life was in some ways an almost closed loop: Objects were reused and by-products were incorporated into new things wherever possible. On the farm, food and plant residues went back into the ground, bottles and other containers were washed and refilled, old clothes were sized down or sewn into quilts, pigs in cities and on farms transformed waste into meat,† while the major urban refuse, manure, was carted off for fertilizer from the mid–nineteenth century onward. Paper and rags always had value—as early as 1690, a mill near Philadelphia made paper from used paper and rags. When cities began collecting garbage in earnest around the turn of the century, many municipal officials shared the conviction of New York City's first commissioner of street cleaning (1895–1898), Colonel George E. Waring, that garbage was a resource.

Known as "the apostle of cleanliness," Waring advocated the separation of wastes to facilitate their reuse and sale. Before World War I more than 70 percent of U.S. cities ran source-separation programs. In 1897 New York City households separated their garbage into three components: ash, putrescible (quickly rotting) garbage, and rubbish. The rubbish was delivered to a "picking yard," where workers separated it into five grades of paper, four grades of metal, and three of carpet, plus bags, twine, rubber, and horsehair.

Many sizable cities like New York pursued the quest for profits

---

*Records on this are, of course, spotty. Waste per capita in this country was estimated to exceed that of Europe in the 1850s, and thereafter.

†In 1925, 44 percent of American cities still used pigs to eat garbage; the practice was outlawed in many places during the 1950s after a series of trichinosis scares. Los Angeles, however, banned swinefeeding only in 1961, and Massachusetts waited until 1968 to do so.

from garbage. They channeled putrescible garbage to rendering plants, where it was steeped and then pressed to squeeze out oils and fats, which were used in making soap, candles, and perfume. Municipal sanitation departments often competed with private, informal scavengers for the revenues from reusable materials, which were seen as a source of financing for public programs. Workers on garbage boats often collected salvageables from the load before it was dumped in the ocean. After a time governments began to charge these "scow trimmers" for the right to skim the garbage.

During the 1920s and through the Depression, the spread of incineration and the dump trucks that carried garbage to cheap land changed the economics of waste disposal. One-stop, comprehensive solutions like burning or dumping were easier and cheaper than separation and processing of wastes. During World War II, urgent shortages of raw materials revived recycling for a time, and saving paper, rags, tin cans, scrap metal, rubber, and even bacon grease became a manifestation of patriotism. Very swiftly thereafter, however, the petrochemical explosion that brought single-use items and convenience packaging into daily life changed the nature of the waste stream and the way people thought about it. Private scavengers persisted for a time in seeking reusable rags, metal, paper, and glass, but with burgeoning consumption and waste, compactor trucks that compressed volume replaced the open trucks in which the scavengers could search for items of value. As Brumberg and Gottlieb have pointed out, around that time garbage ceased to be considered a resource in almost any sense and became merely a material to be discarded.

Although scrap dealers, chiefly in metals, continued to ply their trade where they could, it was only with the advent of the contemporary environmental movement in the late 1960s and early 1970s that the notion of recycling garbage reappeared. At that point it was mostly an idea. Spurred in part by the growing problem of litter, small municipal programs and others run by volunteers focused on newspapers, bottles, and aluminum cans. Although proponents expressed an enthusiasm similar to that of the earlier progressive reformers about "the gold in garbage," recycling levels remained negligible, hovering at around 5 to 7 percent.

The difference between the 1970s and the 1990s is that environmental concern has led to and is now reinforced by serious constraints on waste disposal. When the U.S. Post Office reduced its waste 40 percent through recycling, as its Washington, D.C., headquarters did between October 1990 and January 1992, something serious was afoot.* Beyond the notion that reusing resources will cut disposal costs, the current public commitment to separating and saving, which many thought would spell recycling's doom, itself appears psychically rewarding. Not only are recyclers personally doing something to keep garbage out of dumps and incinerators, but they are also conserving measurable things, like oil, trees, and metals.

Burlington, Vermont's experience of conversion to a recycling approach provides a model of the national shift. As recently as 1986, Vermont's capital city was dumping its trash into an unlined landfill, with no leachate control and no methane collection, next to a wetland. When methane seeped onto nearby properties, residents sued the city and won. But the victory brought more than reparations: It catalyzed a general sense of crisis, spurring wholesale institutional change. As a result, the city established a new public works department and embarked on a farsighted program of waste management geared to maximizing recycling. By July 1993, Burlington's thirty-eight thousand citizens were recycling more than 30 percent of their municipal wastes, through a curbside collection program reaching all households; an "environmental depot" for dropoff of household hazardous wastes, appliances, and tires; a wood and brush depot; a compost project; and (by 1994) the pelletizing of wastewater sludge for fertilizer.† Burlington plans to dispose of 40 percent of its waste through recycling by 1995.

---

*Established as a model for postal service operations nationwide, the Washington, D.C., post office's Saving of America's Resources (SOAR) program source-separates and collects not only high-grade and mixed office paper, newspaper, and corrugated cardboard but also metal soft drink cans, Styrofoam food containers, glass beverage containers, wooden pallets, laser printer cartridges, and scrap metals.

†At the same time, the city sited and built a lined landfill as the environ-

On the other side of the continent, Seattle, Washington, had by 1993 achieved a stunning 42 percent recycling rate on a much larger scale. Measured on a per capita basis, the city of 516,000 has become the "recycling capital of the nation." Like Burlington, Seattle was jolted into action in 1986 when two brimming landfills, which subsequently qualified as Superfund candidates, began leaking methane onto neighboring land. After citizen protests derailed incineration projects in 1985 and 1987, the city contracted with the county government to use their landfill, and committed its future to recycling.

Seattle's vision is laid out in its August 1989 report, *On the Road to Recovery: Seattle's Integrated Solid Waste Management Plan:*

> The scene is Seattle in the year 2010. People are throwing away much less than they did in 1989. They are buying more durable products rather than disposable products. They are buying products with little or no packaging. Most people now choose the half-can garbage collection option and many weeks don't set out a can at all.
>
> Recycling has become a way of life in Seattle. Most homes have only one waste can but every room has a recycle can. Many homes have a compost bin in the back yard. Every home has a special container for food waste. At work, in school, in shopping malls and on street corners, people casually toss their pop cans, bottles, newspapers and paper into recycling containers and their scraps of food into compost buckets. . . .
>
> Processing of recyclables has become a major industry in Seattle, employing hundreds of people. The brightest, most creative young men and women often choose careers in developing new recycling technology and designing new products from recycled materials, using minimal, easily recyclable packaging. . . . These days, so much material is recycled, or never thrown away in the first place, that relatively little is left over that can be called "waste."

---

mentally soundest option for disposal of community wastes while recycling expanded. Comparatively steep (for Vermont) landfill tipping fees of $67 a ton supported the costs of $26 a ton for recycling. Since fees for households are based on the volume of garbage, it pays to recycle. To ensure that success in expanding recycling will not run afoul of a "hungry" landfill, planners sized the facility so its fixed costs would not escalate if volumes fell.

Some may find that this description resembles the northwestern "Ecotopia" described by Ernest Callenbach in his novel of the same name, but if Seattle continues its initiative, it may have a chance of realizing its 60 percent recycling target by 1998.

John E. Pepper, the president of Procter & Gamble, acknowledged in 1991 that his industry was reexamining packaging strategies in order to respond to consumers' negative feelings about waste. He admitted that people's ambivalence about single-use items was forcing the pace on plastics recycling. "Even though consumers like to use plastic packaging, they hate to throw it away," he said. "They feel guilty because plastics are a contributor to our solid-waste problem."

Several families who described their recycling routines to *The New York Times* in 1989 told how it felt to change the ingrained pattern of discard and disappear that to a great extent had insulated them from their waste. Some family members mentioned complying with new trash regulations because they had to at first, but then talked about their current recycling efforts with real enthusiasm. Resenting the need to clutter her kitchen and life with an assortment of bins when local laws mandated recycling, a nurse and mother of three in Hamburg, New York, at first tried to hide recyclables underneath other garbage, stuffing tin cans inside milk cartons and dispersing newspapers among other trash within plastic trash bags thick enough to conceal them. After she realized she was spending more time trying to trick the garbagemen than it would take to obey the rules, she and her husband included a multibin recycling closet in their kitchen remodeling and the whole family adopted new habits of washing and stowing bottles and cans, returning bags to the supermarket, and avoiding Styrofoam. Mellowed by the recycling experience, the former garbage scofflaw commented, "All this does make you think . . . about the future and what it's going to be like" for the children.

A family in Woodbury, New Jersey, was initially bewildered by some of the toughest recycling requirements in the nation, mandating separating trash into eleven receptacles—for colored glass, clear glass, newspaper, mixed paper, cardboard, paper bags, dirty paper, tin cans, aluminum cans, plastic, and nonrecyclable garbage—or risk a $500 fine. Although the family reported that complying with all the regulations took only about ten minutes a day, one family member noted matter-of-factly,

"Every time you're about to throw something out, you have to ask yourself, 'Where does it go?' " Another noted, however, that since they started recycling they had generated almost no regular garbage and said the routine made her feel good, "as though I am making a contribution." The eight-year-old son pointed out the two receptacles in his bedroom (for regular trash and for paper) and proudly demonstrated how he had torn apart the packaging for a new toy in order to toss the plastic wrap and cardboard box in separate baskets.

Like most families in Seattle, this family never gave its trash a thought until the city declared that it was running out of landfill, raised rates for garbage disposal, and announced it would charge by the can. After that, however, the mother of the family, a professional fund-raiser, taught her children which pails were for which garbage and began changing grocery shopping patterns, substituting recyclable aluminum soda cans for large plastic bottles and returning plastic bags to the cashier after her vegetables had been weighed. She summed up both the pragmatic and the cosmic reasons for her concern: "If you hit people in the pocketbook," she said, "it will open their eyes. People don't want to think about garbage. But they're going to have to. It's like pensions." A senior citizen and apartment-dwelling recycler in Brooklyn, New York, made a similar connection between garbage disposal and posterity, asking: "Where are we going to dump all this stuff? The cemeteries?"

For all this good feeling, short-term expectations for recycling have been unrealistically inflated. People want to believe that because it's *right* it will work immediately. In most places, however, high levels of collection and sales will take years to achieve. (By way of comparison, Japan, where land is scarce and recycling has long been culturally ingrained, currently recycles only about 30 percent to 35 percent of household garbage.) When we consider local landfill charges and the market prices for recycled materials, recycling may cost more than dumping for some time to come. Even in Seattle, collection is so far ahead of market development that officials worry about mounting volumes of stored recyclables. Unless public expectations for recycling reflect realities, dashed hopes are likely to erode support.

The current patchwork of both mandatory and voluntary, curbside and dropoff programs offers recycling access to more than 17 million households and thousands of office buildings. Expanding that to reach everyone with regular collection would not be extraordinarily difficult if we were so inclined and were willing to spend the money.* Yet the problem still remains: What to do with all the materials we can collect.

Though the key clearly lies in increasing demand, market development is hardly a straightforward process. Over the past few years it has become evident that amassing a supply of most materials does not automatically trigger demand. Not surprisingly a large supply tends to dampen the price. On the other hand, the recent advent of a collection infrastructure for plastics shows that building a stable supply where none previously existed does bring materials processing closer to reality. Rising collection rates can also drive political action to create a market. Nonetheless, the halting progress made in building markets during the past several years shows the urgent need to change incentives.

The markets for recyclables and reusables have the volatility often endemic to commodities exchanges: They respond according to the economics of distribution and sales, to technology, to direct or indirect pressure that industries feel to incorporate recycling. At one extreme are aluminum and scrap steel, recycled within a thriving commercial market, and at the other, plastics, where enthusiastic collection is driving a reluctant cohort of chemical companies to seek a reverse alchemy they had seen no previous need for. In between are masses of paper and glass of all sorts, rubber (tires)—and the putrescible grass, leaves, and food for compost.† As Susan Williams shows in *Trash to Cash,* her exhaustive investigation of commercial possibilities in the post-consumer waste stream, markets are for the most part still em-

---

*Multiunit apartment buildings present the greatest collection difficulties, and much more experience is needed in building effective programs. Significant success has been shown, however, where financial incentives in buyback programs or savings in collection fees have been brought to bear.

†Although construction and demolition debris (not counted in the municipal solid-waste stream but amounting to at least 25 percent of actual municipal waste totals) will not be discussed here, efforts to divert it from landfills, through waste reduction and reuse, are accelerating.

bryonic, though growing. Moreover, many of the technologies that will bridge the gap between resources and market are mostly in experimental stages or still on the drawing board.

The premier recycling commodity is aluminum. In aluminum can recycling, technology, money, and market all come together. At the time of the first Earth Day in 1970, the nonbiodegradable throwaway aluminum beverage cans that were beginning to dominate soft drink distribution were the major visible targets in the war against litter. Since then they have captured 95 percent of the beverage can market, and are seen as valued assets in the campaign for resource recovery. In recycling aluminum cans, we come close to achieving a closed loop: A can purchased today can be returned to a grocery store as part of a new beverage container in six weeks; it is almost 100 percent recyclable, having a loss of only 5 percent to 15 percent when aluminum scrap is melted.

Recycled cans currently represent about a quarter of the U.S. aluminum industry's raw material supply, and the demand for them seems inexhaustible. Capital costs for melting down used aluminum are 12.5 percent of those for primary smelting, and processing the recycled metal uses 5 percent of the electricity, which accounts for one third of the cost of producing aluminum from virgin materials. With a savings of 7.5 kilowatt-hours of electricity for each pound of recycled aluminum, recycling conserved twelve billion kilowatt-hours in 1989—enough to supply the annual power needs of the city of Boston. For obvious reasons, the aluminum industry buys back every used can it can get its hands on.

Scrap steel has only about a tenth the value of used aluminum, but it has been sold for reuse for over two hundred years and has the highest recycling rate of any material in this country, 66 percent. Scrap, much of it from junked automobiles,* currently

---

*Each year U.S. automakers produce eight million new cars for Americans, and many more are purchased from abroad. Probably at least eight million are junked each year—each one weighing about a ton—and most are shredded, yielding about 80 percent in metal scrap.

supplies about half of the steel consumed by U.S. mills (52.6 million tons or 54 percent in 1989). Most steel products contain about 25 percent recycled material and many are made of 100 percent scrap. Environmental benefits are also high: The steel industry estimates that using recycled materials cuts energy use by three quarters. Annual energy savings from steel recycling add up to six hundred trillion Btu (equivalent to Los Angeles' electric power needs for eight years). According to the EPA, compared to producing steel with virgin materials, using scrap iron and steel brings an 86 percent reduction in air pollution, a 40 percent reduction in water consumption, a 76 percent reduction in water pollution, and a 97 percent reduction in mining wastes.

Paper and paperboard bulks very large in our waste stream, making up 40 percent of all U.S. municipal waste, or six hundred pounds per person per year. It seems benign: After all, it's biodegradable, and most of it currently is landfilled. Yet the decomposition or combustion of paper transfers potent chemicals in inks and processing fluids to air, water, or ash residues. Thus the progress that has been made in recycling paper, particularly newspaper, during the last few years is significant in terms not only of volume (and trees) but of pollution as well. In 1991, recovered paper volumes grew dramatically, by 6.8 percent, to 31.1 million tons.

More than for any other commodity, in the recycling of old newspapers (ONP) public pressure has changed the structure of the industry. Because most newsprint mills are located near the sources of virgin materials—i.e., forests—rather than the urban areas that generate ONP, upping recycled content is a matter not only of washing away inks and clays but of restructuring plant and transport systems. New mills must be located near cities and transport capability between mills and cities intensified. Before 1986, recovery of ONP had remained at 30 percent of output for ten years. After 1988, however, rates jumped, reaching 52.2 percent, 6.5 million tons, in 1991, for a 49 percent increase in those three years. As supplies mounted, prices fell, deflating hopes for the evolution of a viable resale market.

Now, however, an upsurge in state laws mandating recycled content for newspapers is building demand. Under this pressure, the newsprint industry is accelerating investment in de-inking facilities for processing ONP, which when they come on line will increase quantities of old newsprint utilized from 5.4 million tons in 1991 to 6.2 million tons in 1993 to a predicted 8.4 million tons in 1995. When capacity to use ONP more nearly equals supply, as industry experts expect will happen in late 1994, prices ought to rise again. This will give local recycling programs (some of which have been paying to dispose of the collected newspapers) a significant boost.*

Since recycling paper shortens and weakens the cellulose fiber, paper recycling cannot continue in a closed loop indefinitely: A given piece of paper can be recycled perhaps only four to eight times, and most production processes combine virgin and recycled pulp to upgrade quality. Environmental benefits, however, are real. Even though most papers are bleached after de-inking, recycling uses 75 percent less bleach than producing virgin paper, and avoids most of the dioxin formed from the lignin in virgin pulp when it is bleached.† According to the EPA, producing recycled paper results in 75 percent less air pollution and 35 percent less water pollution than virgin milling.

Glass is the only commodity that can be recycled in a truly closed loop, with no loss, no waste, and no by-products. Probably because the virgin materials for glass—sand, limestone, and soda

---

*Recycling levels and markets for other kinds of paper are also growing. Recovery of corrugated containers, for many years the most used paper commodity—sold to commercial collectors by retail stores, supermarkets, and factories—rose 6.8 percent in 1991. Since 1991, technology has enhanced prospects for old magazines, with the development of ways to use the clay coatings that had formerly gummed up the de-inking process. Demand for office waste paper is also expected to rise dramatically in the next several years, doubling between 1993 and 1995, when marketing prospects will improve considerably with the completion of twelve large new de-inking facilities.

Although mixed paper constitutes a significant portion of our waste stream, it is considered in recycling markets the commodity of last resort; consequently recycling levels are very low. A representative of International Paper Co. has suggested a potential use for mixed paper in linerboard, noting that in the United States only 3 percent of linerboard is made from recycled fiber, compared with 54 percent in the rest of the world.

†Although recycling also creates potentially toxic runoffs of inks and clays, the potency of this sludge has diminished significantly as ink manufacturers have reduced the lead, cadmium, chromium, and other heavy metals in ink to comply with EPA regulations for air and water emissions. If resulting sludge tests toxic, it must be disposed of in a hazardous waste facility, but much recycling sludge tests nontoxic and is much in demand by farmers as a clay–heavy soil conditioner.

ash—are so cheap, however, only about 12 percent of all glass—and around 25 percent of glass contrainers—are currently recycled. The principal advantages in recycling are a reduction in air pollution by 20 percent and water pollution by 50 percent—and a hefty energy saving, according to various sources, running anywhere from 30 percent to 45 percent and including mining and transport. As with the newsprint industry, the bottling industry will have to be restructured geographically to make glass recycling work. Except for a group of plants in California and a few in Oklahoma, most glass container plants are clustered in the East. Thus costs mount for transporting crushed glass, or cullet, from all parts of the country.

For plastics, the most widely used material in the United States today, recycling volumes are still minuscule, hovering at around 1.1 percent.* The only plastic item recycled in any quantity is the soda bottle made of polyethylene terephthalate (PET), representing about 5 percent of plastics in the waste stream. These reached a recycling rate of 30 percent in 1990. Coming in a distant second are milk jugs and juice and water containers made of high-density polyethylene (HDPE), representing a much larger share, 21 percent, of the plastics waste stream but a much lower rate of recycling, less than 5 percent.

Starting with collection, but continuing with processing, the fundamental problem for recycling plastics can be summed up in one word—"separation." Plastic products are made of diverse resins derived chiefly from petroleum and natural gas. Many products, like the multilayered squeezable ketchup bottle, are composed of several different resins. Because each has a different melting point and differing reaction to reheating, they cannot be recycled unless they are separated. Melting and mingling them produces a rather blocky end product with very limited applications. Apart from PET and HDPE containers, which are made of a single resin and easily recognized, it is still much easier

---

*As Susan Williams points out in *Trash to Cash,* the recycling movement has, however, already had a significant effect on plastics packaging: in fending off the plastic can. When Coca-Cola and a Swiss firm began test marketing the plastic can in the mid-1980s, environmentalists in concert with the aluminum industry (whose recycling process for aluminum cans was threatened by plastic strays) protested. By 1988, the plastic can appeared to be a dead letter, with nearly two dozen states enacting bans by 1991.

to commingle than to separate mixed resins. Thus, the quintessential product of plastics recycling today remains plastic lumber and other items with limited market potential like plastic park benches, flowerpots, trash cans, indoor-outdoor carpet, or fiberfill.

In fact, however, the obstacles to plastics recycling are not mainly technical, for much of the technology necessary to close the loop already exists.* It is both infrastructure and market incentives that must be developed. Plastics recycling plants can be built fairly quickly, in around ten months, versus three years for a recycled-paper mill. Because plastic need only be remelted in order to be reused, recycling can save 92 percent to 98 percent of the energy expenditure required with virgin materials. Recycling facilities emit far fewer pollutants, which in plastics manufacturing are frequently highly flammable and toxic.

In spite of the technical and even economic advantages, a meaningful effort to achieve plastics recycling is unlikely without major changes in economic incentives for structural change. Neither the 25 percent recycling level promised by the plastics industry by 1995 nor the higher levels that would represent meaningful reduction of waste seem imminent. Except for PET and HDPE, real market advantage awaits the development of collection and processing infrastructures that do not yet exist. Incentives for the development of this infrastructure are nowhere on the horizon. Making plastics recycling viable will entail changing the way we figure costs, to include avoided costs of waste disposal and pollution mitigation as well as energy savings, on the revenue side of the balance sheet.

One specific type of recycling, yard waste composting, is sweeping the country. Not surprisingly, America's lawn-care mania generates some thirty million tons of grass, leaves, and brush (around 19 percent of the municipal solid-waste stream) every year. An idea whose time has come, laws banning yard wastes from landfills were passed in fifteen states between July 1990 and February 1992. Since then, more states have passed such laws.

When this vegetation breaks down into a rich soil additive, will we be heading toward a compost glut? Although composted yard

---

*For a further discussion of plastics recycling technologies, see Chapter 8, pages 181–183.

waste does not contain the same intense mix of nutrients as fertilizer,* as a soil additive it aids moisture and nutrient retention,
decreases erosion, improves drainage, and suppresses plant disease. The potential agricultural market for this mulch should be
enormous. In this country alone, an estimated twenty-five million
acres of severely eroded crop land loses about fifteen tons of soil
per acre a year; restoring an inch of topsoil takes sixty-five tons
of compost. And the going rate for composted yard waste of zero
to twenty-five dollars per ton should make it a bargain. (The cost
per ton for collection ranges from zero to eighty dollars and for
processing from four dollars to twenty-three dollars.) Again, however, closing the loop from dirt to dirt will undoubtedly require
legislation to change incentives for farmers and agribusinesses.

The current public and official focus on recycling largely side-
steps source reduction and reuse, which are likely to be a much
more difficult proposition. Recycling has already brought about
real shifts in people's habits and some adjustments in manufacturing processes that support waste reduction generally. Source
reduction will, however, require much more radical changes, affecting industrial design, consumption patterns, and the way the
economy works.

Moving toward source reduction will require a sizable political
and imaginative leap. One main aim of source reduction, eliminating *polluting* materials from both processes and products, is
now widely accepted as "pollution prevention." The overall goal
is, however, much broader—and more threatening to the economic and political status quo. Recycling takes products, processes them, and manufactures new items out of the constituent
materials; source reduction would cut down on the volume of
materials in manufacturing and even the volume of things manufactured. It works to maximize the efficiency of materials use in
the manufacturing process and fosters design of products that
can be used again and again. Looking at a computer, for example, source reduction would mean process changes like cut-

---

*Mixed food and yard wastes yield a richer compost—the higher the pro-
portion of food wastes the more nutrients.

ting back solvents in fabrication of its circuitry, streamlining packaging and requiring its return, and designing machines with replaceable parts to allow continual updating. If it were successful, source reduction would eliminate the need for some manufacturing and consequently might seem to work against "growing" productivity.

An analogous problem in pushing for source reduction by consumers is that preventing waste often means not buying a product. As environmental economist Bette Fishbein has written, because "preventing the generation of garbage implies facilities that are not built and materials that are not collected, not marketed, not sold . . . it seems in many ways like a nonevent." While it makes sense for industry to economize on materials use or, in response to regulations, to cut pollution, for consumers, source reduction may require reassertion of old-fashioned thrift in a marketplace that discourages it.

Consumers' efforts will entail changing habits of buying and use, both to cut down on trash and to put pressure on manufacturers. Switching from disposables to reusables, continuing to use things even after the first gloss has worn off, rejecting overpackaged products, buying in bulk, and looking for nontoxic alternatives and appliances that don't run on batteries are some options for consumers who care about the impact of their consumption patterns on the environment. As with recycling, much of what people can do personally amounts to a sort of individual war on throwaway packaging and single-use items—toting canvas carry-alls to market, reusing plastic bags, buying in bulk and refilling plastic containers, disposing of razor blades, not razors, using cloth towels and napkins the way grandmother did, and registering with the post office an unwillingness to receive junk mail.

Large-scale purchasers like stores, restaurants, business, industry, and government, given their size, can eliminate huge chunks of postconsumer waste, particularly from the fast-growing paper stream.* In addition to recycling business paper, offices can save money and avoid waste-disposal charges by simple changes in behavior and technology, like using all paper (even in printing

---

*Office paper increased from 1.7 percent of the paper stream in 1960 to 4.1 percent in 1988, and is projected to be 6.4 percent in 2010.

and photocopying) on two sides, sharing and circulating documents, faxing directly from computers, shipping all materials in reusable containers, and contracting with suppliers who do so. In businesses as in the home, returning to reusables such as tableware, towels, and refillable laser cartridges, pens, and tape dispensers, as well as choosing equipment for its durability and long-term service warranties, substitutes time for materials.

In a key aspect of source reduction, reuse, the United States has particular problems. In its headlong rush to single-use containers after the 1950s, the United States differs from most other countries, where refillables are still the norm. The reasons have more to do with strategies for market dominance within the bottling industry than with economic costs or environmental benefits, both of which favor refillables. Although a number of countries, most notably Canada, Germany, the Netherlands, and Denmark, mandate the use of refillables, very little legislation promoting them has passed in the United States.

In the longer run, however, we can only reduce waste effectively through the use of fewer objects. The pursuit of long-lived and reusable goods can probably best be achieved through a new sort of link between producers and consumers—through service industries that take charge of repair, reconditioning, and technical upgrading. As Swiss economist Walter R. Stahel explains, in this approach the purveyors would not sell the goods themselves, but would market the utilization of the goods. Correspondingly, the consumer would purchase not the object but its use. In order to get the highest return on their capital investment, it would behoove purveyors to provide durable goods and cost-effective maintenance.

Ultimately, making postconsumer waste and repair the responsibility of the vendor is the most effective way to achieve waste reduction. This is happening now in Germany, where government legislation is forcing manufacturers to "take back" packaging wastes and ultimately discarded products as well.* When "take-back" regulations take effect, self-interest dictates more efficient use of materials and energy. Again, changes in market incentives to raise costs of materials (which we will look at in

---

*For a fuller description of Germany's "take back" program, see Chapter 8, pages 199–201.

later chapters) could increase advantages of combining manufacturing and leasing operations. When this happened, a pur-
veyor of automobiles would have an incentive to design upgrades
for existing cars rather than looking for innovations that would
ensure the obsolesence of current models. Again, if farmers purchased security from pests rather than buying pesticides, the pur-
veyor of such a service would have every reason to apply chemical
and biological substances as sparingly as possible.*

Because it upends our notions about ownership of property
and about production, shifting to production and consumption
patterns that really reduced waste would require a massive
change in incentives. We need to make that shift. At this point
the growing costs of reacting to pollution with command and
control regulations and cleanup threaten to swamp our commitment to environmental protection. If we want to ensure that the
sort of environmental mop-up represented by Superfund is a
one-time cost rather than an ongoing drain, we will have to stop
producing the waste—much of it toxic. We need to change macroeconomic policies to make waste unprofitable. Can we do that?

---

*For further discussion of an economic approach replacing consumption
with utilization, see Chapter 13, pages 286–291.

Part III

Reducing Waste

8

≈≈≈≈≈≈≈≈

Can Technology Fix It?

Nature is no great mother who has borne us. She is our creation. It is in our brain that she quickens to life.

—OSCAR WILDE, *The Decay of Lying*

You aint heard nothin' yet, folks.

—AL JOLSON in *The Jazz Singer*

A MERICANS WILL FIERCELY resist the idea that technology and the market cannot get us through this crisis. Given our strong national preference for the workings of the invisible hand, the notion that if we just get out of the way, commercial "progress" will make our lives better continues to hold considerable power. We are inclined to forget that technology and the market are only tools.

Our embrace of recycling itself plays directly to our twin faiths in technology and commerce. Reusing materials instead of burying them makes sense, and, given enough human ingenuity and effort, might even make money. The idea that a new manufacturing process may turn what were useless items into salable commodities reflects the optimism and pragmatism that has traditionally underlain our economic growth and innovation.

Technology and the market, in fact, are ultimately the best tools we have to cope with waste. As things stand, however, reliance on either—or even on both together—is a recipe for continued environmental erosion. Only if government sets a new framework for using our formidable technical skills and commercial prowess can we use them to do the job.

Admittedly the past several decades have strained our passion for technology considerably. As we have seen, optimism about the seemingly inexorable advances of science and industry has become muted in the wake of progressively revealed damage from pollution and industrial accidents. For the first time since the Enlightenment, doubt that we can shape and control our impact on nature is widespread. The NIMBY phenomenon, and particularly the resistance to nuclear energy and to incineration, shows the extent of people's skepticism about new machines and experts. Sociologist David Lowenthal has remarked, "Today we dread both what we know and what remains unknown."

We have come to suspect that a technology's effects usually will surprise us, that its unintended consequences frequently will loom larger than those we foresaw, and that predictions about "modernistic" futures often miss the technological and scientific phenomena that most change our lives—like birth control, fax machines, and McDonald's. Technology makes us rich but the experience leaves us torn. We no longer feel the joy and wonder of our forebears as the lights went on, people moved on a screen, and voices came from a box. We are beginning to anticipate that science will illuminate a problem like global warming more readily than it will reveal genuine solutions for that problem.

Yet long-held ideas do not loose their grip easily. Although many Americans may now believe there are no easy answers, we can still imagine a technical fix much more readily than a social one. Evidence of technology's power is everywhere. Where Vietnam assaulted our faith in our high-tech military prowess, Desert Storm showed the utility of our huge investment in state-of-the-art arms and materiel: Because we had the goods, we were invincible. Medical research maintains life in premature infants through microscopic reconstruction of their tiny organs; next may come embryonic gene-splicing to forestall birth defects. Computers fly planes, design houses, track and analyze the buying habits of millions.

For all our skepticism, we still harbor a lot of hope that when a new problem is perceived, the technology to solve it will also be found. After the polio epidemic of the 1950s, Dr. Jonas Salk invented a vaccine; when graffiti artists defaced New York's subways, spray-proof paint was developed; now, as traffic clogs our major arteries, we look forward to "smart roads" that will redirect the flow and assess charges for use. Particularly as we doubt our own or government's ability to produce societal progress, we continue to place relatively more credence in technological progress and the products of individual ingenuity.

In the environmental area, technological progress in fact gives reason for hope. In greater profusion than ever, new products and processes pour from factories, laboratories, and people's garages. From the two-way envelope to microbes that eat toxic materials to an automobile designed for disassembly, inventions are emerging that aim to reduce waste.*

Most new technologies—for recycling and composting, curtailing agricultural runoffs and oil spills, and facilitating municipal and industrial waste reduction—are not yet at the stage of commercial development. Because many of them threaten the existing comparative advantage of established industries, they will not soon achieve economic viability without major shifts in policy. Nonetheless, it is worth exploring the incredible profusion of promising technologies[†] already in view that can help us solve our waste crisis, in order to underline an important reality: The barriers to coping with our crisis are not technical but political.

---

*Though they are not discussed here, advances in energy technologies are also highly relevant to waste prevention and materials management: In a postcarbon world, the development of abundant nonpolluting energy sources will greatly increase our flexibility in reusing, reclaiming, recycling, and detoxifying materials. Energy is also needed for various kinds of catalytic reactions that destroy polluting chemicals in water. Although both recycling and reuse in most cases save energy, the trade-offs, involving transportation and cleaning, among other costs, are complex.

†Sometimes, of course, human ingenuity appears to go fairly far awry, as in a Japanese scheme for greening the Sahara by injecting it with vast volumes of water-retaining polymer gel, Mexico City's scheme to blow pollution away with giant fans, or the offer by a private Russian company to vaporize toxic waste in underground nuclear explosions. Another megascheme with similar earmarks, currently being considered by U.S. scientists, involves trapping ozone-depleting carbon dioxide in the oceans by fertilizing microscopic plants in the waters around Antarctica, growing millions of miles of CO_2-consuming seaweed on floats, and pumping billions of pounds of CO_2 into ocean depths.

∾

Reading garbage industry magazines like *Waste Age* and environmental journals like *Biocycle* and *Resource Recycling*, as well as a raft of new periodicals aimed at the business of environmentalism and environmentalism in business, makes it evident that better mousetraps abound and that a great deal of effort is being invested in building them. Our fledgling attempts to revive recycling in the late twentieth century engendered experimentation with a rash of ideas. Because collection and separation are local and mostly small-scale, communities have tailored their systems to their individual waste streams and budgets. After debating the merits of various sorts of buckets and bins, their fit with one- or two-person compartmentalized trucks, and the need for compactors or automatic lifts, towns opt for widely varying programs.

Collection and separation facilities, known as Materials Recovery Facilities or MRFs (pronounced "murfs"), are highly labor-intensive. They also embody a customized assemblage of computerized scales, conveyor belts and chutes, hammermills and shredders, air-stream classifiers, magnetic separators, trommel screens, glass crushers and grinders, balers, chemical baths, and foam flotation. Practical and ingenious rather than high-tech, these facilities embody the vision of their individual proprietors about how to solve a very concrete set of problems. Not surprisingly, the result sometimes looks more like a Rube Goldberg device than late-twentieth-century American industrial technology.

Among gadgets facilitating easier collection of bottles and cans are a bucket-sized glass crusher, designed for bars, that turns piles of empties into salable cullet, and reverse vending machines. The latter are already gobbling, magnetically sorting, and crushing empty cans in a number of U.S. supermarkets. Several variants dispense supermarket coupons as well as change. A Swiss version, called Egapro, infuses the virtuous process of recycling with the seductions of gambling. The device operates like a slot machine, with a row of wheels on the face of the machine spinning off occasional jackpots—in the form of game tickets entitling the player to prizes like coupons, raffle tickets, or meals at local fast-food restaurants—in exchange for cans.

New technologies are also changing recycling. With some materials, notably paper, the process depends on stripping away extraneous elements. Until recently, the clay coatings on slick paper made magazines and catalogs highly undesirable commodities. With the discovery in 1990 that the clay coatings could aid flotation of paper through de-inking, a whole new category of paper entered the recycling stream.

For that most ingenious of alchemical puzzles—that is, plastic—the difficulties, as we have seen, lie mainly in separating constitutent components. Many plastics manufacturers would prefer to finesse the problem by opting for a much simpler technical fix: melting down plastic products into fuel. German manufacturers of plastic packaging, spurred by their country's legislation requiring them to "take back" their wares after use, are pushing the reuse of resins as energy. Because the process itself uses up considerable energy and creates significant quantities of toxic by-products, it appears in reality to be a fancy version of incineration.

In order to really close the loop, recyclers must separate resins. A number of different separation technologies are evolving very quickly, though most are at the experimental stage. The system of identification for single-use plastic containers developed by the Society of the Plastics Industry helps at the outset. The familiar triangulated recycling arrows accompanied by a numerical resin symbol identify six main resins for easier manual sorting.

Most sorting is now done by hand, but the need for greater rigor and efficiency is driving the system toward automation. The pioneer in automated sorting is Dr. Henry Frankel of Rutgers University's Center for Plastic Recycling Research, who developed and sold the first mechanical system for separating PET (polyethylene terephthalate) and PVC (polyvinyl chloride) containers in 1989, using the light transmission properties of the resins. Since then, Frankel has evolved a single-station, multidetector system with computer software to eject containers from a conveyor belt according to resin and color. The Rutgers system sold for $35,000–$45,000 in 1992; eight models of a similar though larger system manufactured by National Recovery Technologies of Nashville, Tennessee (priced under $200,000), were sold during 1990 and 1991. Further iterations, mostly in demonstration stages but destined to sell for around $350,000 to $750,000, separate containers into three, four, and seven categories.

Eastman Kodak has contrived a way of sorting both resin categories and different grades of the same resin by scanning under ultraviolet light for organic marker compounds incorporated into the virgin resins. (Obviously its use would require industry-wide cooperation, of a sort that does not now exist, in inserting the markers.)

Other inventions may help lower transport costs for bulky plastic containers. The beneficial impact on costs of a new baling system developed by the Council for Solid Waste Solutions demonstrates the importance of plastics bulk to the bottom line: After the Milford, New Hampshire, MRF installed the system, which perforates and compresses plastic bottles into bales, transport costs per mile decreased 60 percent and revenues increased 91 percent per pound as a result of the consistent weight, uniformity, and quality of the bale. Facilitating the separation of plastics after they have been shredded would eliminate further bottlenecks because shredding the containers significantly reduces volume during transport. Commingled shredded plastics can be sorted by flotation and air pressure as well as by centrifugal force, which segregates materials of different densities. DevTech Laboratories of Amherst, New Hampshire, has developed a two-hundred-pph (pounds per hour) prototype machine for separation of PET and PVC particles on the basis of their electrostactic properties.

In addition, several chemical processes for separating molecules now being developed will represent giant strides toward closing the loop. Goodyear and Du Pont are already recycling PET by applying heat and pressure to the plastics in the presence of chemicals such as methanol. This causes the chains of molecules that make up plastic resins, called polymers, to break down into their constituent monomers, which can then be recombined into virgin resins. A second method, still wholly experimental, recovers the various different polymers (which dissolve at different temperatures) from commingled plastics by immersing them in a solvent and recovering the polymers one at a time from the solvent as the temperature is raised and the solvent repeatedly evaporated.

Batelle Memorial Institute is developing a reverse polymerization process that goes further. Reportedly still years away from commercial use, the system reduces mixed plastics in a thermal reactor into gases consisting of monomers that can then be separated and used to produce new plastic resins. One awed plastics expert remarked that this would be "the ultimate recycling technology, tantamount to recycling glass back into sand, limestone, and soda ash, or paper back to trees."

Researchers are also trying another tack entirely, endeavoring to make mixed resins more useful. Dow and Du Pont are working on developing molecules that can bind together unlike resins into a new chain with new applications. In addition, General Electric is experimenting with a "universal compatibilizer" that could fit several different polymers into its chemical structure.

Many of the new technologies for waste disposal or reduction reflect a quest for natural processes that can do the job less intrusively than through massive excavation, incineration, or chemical treatment, by reclaiming materials or breaking them down into their basic elements. PCBs, for example, may be susceptible to a dose of lime mixed with other trace metals; molten metals can also be used to break down the chemical bonds in PCBs into hydrogen, carbon, and chlorine, which can be reused. At New England's only cement factory, the Dragon Products Company of Thomaston, Maine, an enterprising geochemist devised a "recovery scrubber" that turns pollutants back into raw materials, combining sulfur dioxide emissions with waste dust and ash to produce fertilizer, distilled water, and limestone. For cleaning up oil spills, glass beads coated with titanium oxide that oxidize hydrocarbons when exposed to sunlight, and chemicals that gather up oil into a viscous mass, appear to offer attractive alternatives to the chemical dispersants, themselves often polluting, currently preferred by the oil companies.* Hospitals may soon be able to forgo incinerating polluting, infectious trash, instead subjecting their wastes to treatment, by electron beams or plasma arc

---

*Not surprisingly, given the nature of our nuclear dilemmas, schemes to "transmute" nuclear wastes—largely involving the mitigation of long-term dangers by the separation of elements—arouse a good deal of controversy. Microbes, clay, and the bombardment of wastes with neutrons and chemical separation all have their proponents. All would leave considerable radioactive residues (in some cases increasing the volume of wastes).

torches, which can sometimes transform the wastes into glass or sand.

Many of the most promising technologies reflect an increasing reliance on biology rather than chemistry, in effect reversing the direction of the past five centuries. In that period we moved away from biology toward chemistry, toward the machine, from natural materials to synthetics, from wood to plastics. Now, however, many of the most interesting possibilities for creating new materials, fuels, and foods and for analyzing and manipulating genes, microorganisms, plants, and animals arise from work in biotechnology as well as from integrated efforts in biology, physics, and chemistry aimed at finding new ways to use natural materials. After many years in thrall to the miracles of chemistry and physics, we are beginning to suspect that when biological and chemical processes compete to achieve the same end, biology may often be simpler, safer, and cheaper.

Working through the agency of bacteria may still sound vaguely unhealthy or unsafe. Humans have been creating new possibilities through biotechnology for thousands of years, ever since they learned that yeast would raise bread and ferment beer. During the mid–nineteenth century, however, the idea that fighting disease means eradicating germs gave bacteria a bad name. Although we've long known that we must coexist with bacteria and other microorganisms, since the beginning of this century we've tried to keep the upper hand through our mastery of chemical antidotes. It has taken some reshuffling of our mental imagery to include the notion of these organisms working energetically on our behalf. Perhaps "microbe" has a more positive sound than "bacteria."

Composting is a form of biological recycling. Arguably the most down-to-earth example of biotechnology at work on waste is the handy vermicomposting system—or worm bin. Appropriate for home composting, one pound of worms, bedded down on shredded newspapers in a bench-sized box, can reduce a family's food scraps by about 50 percent. With periodic rotations of the compost in progress with new garbage and paper, the transformation process takes six to eight months and produces a rich soil amendment.

On a larger scale, composting itself is taking on some aspects of high technology. Composting devices used successfully for some years in Europe are being tested by American cities; for example, the Sorain "Mini-Digester," tried in Santa Barbara, which provides a controlled aerobic decomposition and stabilization process employing agitation and air injection within a cylindrical vessel. In the hands of master composter William Brinton, on the other hand, the technology appears subordinate to something like gastronomy. Described by *The Wall Street Journal* as "the Julia Child of garbage," Brinton and the staff of his Woods End Research Laboratory in Mount Vernon, Maine, use computers to calculate the right ingredients for their compost mix, machines to analyze plant nutrients in the compost, and yard-long thermometers to check its heat, endlessly experimenting to get the right recipe. Working for clients including Disney World and New York City, they also trouble-shoot garbage emergencies like disposing of sixteen thousand tons of diseased potatoes from a blight on Prince Edward Island, turning to mulch chickens smothered in a poultry farm's fire, even breaking down, at considerably lower cost than conventional means of disposal, explosive sludge contaminated with TNT.

Biology offers attractive options for sewage treatment as well. Microorganisms break down human wastes through the enhancement of naturally occurring human bacteria. Now, however, adding plants and their attendant microbes to the sewage treatment process may provide a relatively inexpensive way out of the sewage disposal impasse for many communities. Two decades and more than $70 billion after the passage of the Clean Water Act, the effluent flowing out of the majority of the nation's 15,348 sewage plants, after primary and often also secondary treatment to remove solids and restore oxygen, still contains quantities of oxygen-depleting nitrogen and phosphorus, as well as toxic chemicals and metals. Tertiary treatment, conventionally employing chemical precipitation using aluminum, is very expensive and leaves its own toxic residues.

Biological water treatment, on the other hand, offers several options geared to towns and cities of different sizes and locales.

In aquatic-plant treatment systems, the roots of plants that grow in water or in swampy soil, gravel, or rock ponds, such as water hyacinths, pennywort and bulrushes, harbor bacteria and other microorganisms that feed on the minerals and organic matter in sewage, and in turn produce sugars and amino acids to feed the plants. This symbiotic process saves energy by dispensing with the compressors, mixers, aerators and diffusers employed by conventional state-of-the-art sewage technology and, unlike conventional treatment, creates no new sludge. Clearly, aquatic systems are attractive in environmental and budgetary terms and are in fact beautiful.

Only in the early 1970s did people begin to think about the use of plants to purify sewage, but today, man-made wetlands and greenhouses are successfully cleansing sewage effluents in a number of U.S. communities. One of the pioneers of this method was NASA engineer Bill Wolverton, who discovered the filtration capabilities of the Florida wetlands while looking for defenses against biological warfare. In the late seventies he built the first aquatic-plant treatment system in the United States for the Space Lab. In Arcata, California, a fishing and timber town about three hundred miles north of San Francisco, with a population of fifteen thousand, and Denham Springs, in southwest Louisiana, with a population of twenty thousand, successful town marsh treatment systems attract environmental tourists and would-be imitators.

Relying on acres of marshes or hothouse tanks and plants to handle noxious wastes, however, takes a new sort of leap of technological faith. Arcata's mayor and public-works director took that leap in the mid-seventies when they decided to pass up federal dollars available for a regional sewage treatment plant and construct one of the nation's first marsh-ecology treatment systems. After ten years of politics, planning, and planting, and an investment of $5 million—$7 million less than for a conventional plant—in 1986 the city started circulating sewage through the marshes. Today the town's sewage flows out of the primary treatment plant, which screens it and lets solids settle, through fifty acres of cattails and bulrushes, where woodchip trails, feeding egrets, ducks, and sandpipers attract hikers and birdwatchers. When it enters the Pacific at Humboldt Bay it is cleaner than the bay's waters.

Located inland, Denham Springs recently constructed a mammoth rock marsh to do the same job. The town courses its partially treated sewage through three shallow five-acre pits that are lined with stones, filled with water, and planted with canna lilies. Designed to treat three million gallons of sewage a day, the fields of lilies amid stones look like an American variant of a Japanese garden, adapted to our open spaces.

Testing a version applicable to larger cities and cold climates, the solar-aquatics technology developed in Providence, Rhode Island, by John Todd of Cape Cod's New Alchemy Institute works within the confines of a 30- by 120-foot greenhouse. Todd's Rhode Island pilot project, adjacent to a wastewater treatment facility, purifies up to twenty thousand gallons a day by circulating the effluent through transparent plastic silos, each an engineered ecosystem supporting different mixes of bacteria, zooplankton, algae, plants, snails, fish, and crayfish; as well as into engineered marshes via plastic troughs and metal washtubs. Todd's projects in Rhode Island and Vermont, and another pilot plant in Ithaca, New York, appear to show considerable economic advantages over conventional treatment, purportedly taking up no more space than a conventional plant and costing, according to Todd, only two thirds as much to start up and run. Some solar-aquatics enthusiasts have adapted the system for home use. Bill Wolverton designed a system to flush sewage from the bathroom into a plant filter in the Florida room of his Picayune, Mississippi, home.

Now that we are getting the bugs on target, microbial degradation of petrochemical waste and oil spills also shows promise.* Using bacteria, fungi, or other microorganisms to infiltrate these wastes rather than excavating or incinerating them, biological treatment often costs less than a third as much as incineration. Recent research and experimentation have moved from the simple aeration and fertilization of bacteria in toxic soils to stimulating microbial growth and to identification of toxin-destroying genes and their selective breeding for a particular task.†

---

*Bacterial treatments cannot degrade metals into innocuous substances as they do petrochemicals, but only transform them into other chemical species of the metal that may be more or less mobile in the environment. Many solvents derived from chlorine also resist these treatments—they require degradation in an anaerobic (oxygen-deprived) environment.

†Engineering new microorganisms through the manipulation of genetic ma-

After testing microbial agents in part of a chemical-filled lagoon in Houston, Texas, in 1988, engineers estimated that microbial cleanup would cost about a third as much as incineration and take perhaps half as long to complete; in 1991, General Electric replaced an expensive and cumbersome distillation process designed to recover the suspected carcinogen methylene chloride with microbial breakdown into water, carbon dioxide, and salt; in 1992, researchers at a wood treatment plant in Brookhaven, Mississippi, discovered a way to further clean up after the cleanup of the toxic wood preservative pentachlorophenol (PCP), by successfully pitting white-rot fungi against the tough compound left after bacterial enzymes had done what they could. In late 1993, according to the EPA, officials at 159 hazardous-waste sites were considering, planning, or using full-scale bioremediation. Biological treatment is also being used increasingly in Superfund cleanups.

In agriculture, biology also appears to be making a comeback after fifty years in the shadow of chemical technologies. In 1990 only about 1 percent of the $20 billion spent globally to fight agricultural pests went for biological controls. Nonetheless, research trends pointed in a different direction. In 1992, about 30 percent of the five hundred or so scientific papers presented at the annual conference of plant pathologists focused on biological controls, compared to around 1 percent in 1980. In addition, the EPA reports that half of all field research applications from pesticide manufacturers now involve biological controls, representing extraordinary growth over the last decade.

From insect predators marketed in bulk to herbicide-resistant plants to genetically engineered bacteria, the abundant array of biotechnical possibilities reflects both the complexity of the microbial life within soils and the ingenuity of human science. Going beyond the mass-produced insect and parasitic predators that more and more farmers now use, recent research has led, for example, to the development of a line of salts from naturally derived fatty acids that kill insects, weeds, mosses, and lichens; to the use of bacteria to help kill weeds that are sprayed with diluted

---

terial seems more problematic, both because of public resistance and because of the difficulty of keeping them alive under field conditions.

concentrations of chemical herbicides; to the production of minute capsules encasing dead bacteria genetically engineered to manufacture a toxin; and to the insertion of special poison-making genes into crops like tobacco, tomatoes, and cotton to discourage predators.*

Alternative agricultural methods take a variety of forms. Advocates of "biological" approaches range along a spectrum, from proponents of letting nature do the job to those who favor extensive manipulation. The question posed by prairie ecologist Wes Jackson, founder of the Land Institute in Salina, Kansas, sums up the hands-off perspective: Will our society respond to environmental overload with "more heroic cleverness," or rather "a stepping-back to say, 'What has nature been doing here?'" Jackson's own work is aimed at producing grain in a field of mixed crops, with minimal cultivation and no chemical inputs, emulating the natural cycles of the prairie. His approach is diametrically different from that of agribusiness, as well as from that of many of the new biotech companies that he feels too often ignore the complexity of biological interactions while they search for the silver bullet.

In between are the American farmers who practice various forms of alternative agriculture, including the 1 percent to 2 percent identified as organic growers who eschew chemicals entirely or almost entirely, and an estimated 30 percent to 40 percent who have begun to seek ways to escape the escalating cycles of chemical dependence through lowering fertilizer inputs. With traditional "technologies" like crop rotation, ridge tilling, and plowing under of winter cover crops, they are enlisting nature as an ally in saving energy, fertilizer, and pesticides as well as in conserving soils. As a result, the costly chemical spiral is slowing, with pesticide use declining since 1982, and fertilizer use also somewhat lower in 1992 than in 1980.

---

*Other promising technologies include infecting a cotton-eating caterpillar with a virus engineered from an enzyme that stops the caterpillar from feeding, and splicing together killer viruses with toxin-producing genes that paralyze insects to keep them from eating crops while they are dying.

Policy lags seriously behind technology, however. Alternative technologies get only minuscule amounts of research assistance. In 1992 only about 1 percent of the Department of Agriculture's research budget went to the department's Low Input Sustainable Agriculture (LISA) program.*

For farmers, the choices are also limited by a number of entrenched institutional obstacles to deviating from established practices. The known advantages of chemicals count heavily. Chemical controls can wipe out a broad array of pests *immediately,* whereas biopesticides aim more narrowly, take longer, and still seem less effective against weeds and some fungi. Moreover, biotechnology for agriculture runs headlong into a well-entrenched and formidable competitor in the chemical industry.

Although agribusiness corporations are funding biological research, the main business of agribusiness is still selling chemicals. Firmly lined up behind the industry's stake in $14 billion annual sales of chemical fertilizers and pesticides, biological research funded by agribusiness places heavy emphasis on biological means to make chemistry work better, for example, by increasing plant uptake of fertilizer through a nitrogen-capturing microbe attached to seeds, and on time-release fertilizers and pesticides that make it possible to target inputs more precisely. The industry has invested most extensively in ways to make plants more compatible with chemicals by increasing their resistance to insecticides and herbicides. Experimentation at land grant universities, which derives substantial funding from agribusiness, also reflects these industry priorities.

Like agribusiness, the petroleum industry appears to have little interest in hastening a pending biotech revolution in plastics manufacture. Plastics fabricated synthetically from petroleum do

---

*LISA's purview includes comparisons of low input with traditional methods (e.g., yields of strawberry growers who kill pests by covering the soil with plastic versus those who use pesticides) and ways to ration chemicals more precisely. In 1990–91, LISA's budget was only $6.3 million, within an overall DOA research allocation of $500 million. Chemical lobbies like the National Agricultural Chemicals Association resist the idea of alternative agriculture—and the erosion of huge fertilizer and pesticides sales—fiercely. In Iowa, for example, crowds of chemical dealers besieged the state legislature to fight a bill creating a Center for Sustainable Agriculture at Iowa State University funded by agrichemical sales taxes; when the bill passed, the industry targeted for defeat legislators who voted for it.

not biodegrade because bacteria and fungi are equipped with enzymes to metabolize only molecules produced by nature. Making completely biodegradable plastics from plants, however, is possible now. After extensive research to assure their stability in damp environments, a variety of natural plastics suitable for an array of uses and later composting can now be produced. One promising process depends on a bacterium, the *Alcaligenes eutrophus*, which manufactures a natural polymer to store energy in much the same way we accumulate fat. Discovered by a biologist at the Pasteur Institute in Paris in 1926, the bacterium converts sugars and starches into polymers. These natural plastics have a strength and durability comparable to those of petroleum-based synthetics in ordinary use and can be melted, molded, and shaped into any form. In 1992, a British company using these bacteria at a pilot plant was producing biopolymers for about twelve dollars a pound and expected the price to fall below five dollars by the mid-1990s (versus thirty to sixty cents for common plastics).

Researchers at Michigan State University and James Madison University in Harrisonburg, Virginia, have also succeeded in producing plastic *within* plants. In 1992 they reported success in efforts to alter the genes of a small weed related to the mustard family so that it would produce a plastic similar to polypropylene. They plan to work next on altering the same genes in major starch producers like the potato and the sugar beet. Batelle Memorial Institute has developed yet another process using bacteria and chemical processes to convert blackstrap molasses into acrylates used in acrylic paints. In addition, at a plant in Rockford, Illinois, the Warner-Lambert Company is manufacturing natural plastics from starch made directly from corn or potatoes—without bacteria or genetic engineering—by melting it under high pressure in the presence of water.*

---

*A substance derived from the shells of crustaceans and the walls of fungi (called chitin) also has promise for possible uses including food packaging, fabrics, and films, as well as for binding metal ions in water purification. An African plant, the fibrous kenaf, may become an environmentally superior feedstock for paper pulp, producing three to five times more pulp per acre than trees do, at roughly half the cost, and requiring fewer chemicals to whiten fibers for paper. Its fibers may also be useful for molded fiber parts for automobiles, roofing, felt, fire logs, carpet backing and passing, and cattle feed.

Getting these natural plastics to market is likely to be much harder than producing them, however. That will depend almost entirely on the price of oil.* Our cheap energy policy stands athwart our innovation process, here as elsewhere.

Looking at materials use generally, the consumption of both energy and materials has been declining in relation to the growth of the economy. Between 1978 and 1987 U.S. energy consumption per household dropped from 138 million Btu to 101 million Btu. The U.S. economy is 30 percent more energy-efficient than it was two decades ago. Part of this reduction stems from a decline in the consumption, and particularly the processing, of basic industrial materials such as steel, copper, and cement. Since the 1970s, the use of bulk materials, including paper, fertilizers, synthetic fibers, other industrial chemicals, cement, glass, pottery, and metals, has also fallen, though less rapidly than the basics, in relation to total industrial production. (The only bulk materials whose consumption is still growing as fast as or faster than the economy are plastics and industrial gases.)

This phenomenon, known as "dematerialization," has resulted not from efforts at conservation, but from the effect of rising energy costs on materials processing, the availability of lighter substitute materials, changes in production technologies, and market saturation. Demand for steel, for example, had begun to fall in relation to GNP by 1920, and for steel-based consumer goods, in relation to population, by the 1950s. An office building that required a hundred thousand tons of steel thirty years ago can be built with thirty-five thousand tons today. By now the domestic market for durable consumer goods having a high content of materials per dollar, such as cars and appliances, is largely glutted. In response to advances in materials substitution, many consumer goods, such as cars, television sets, irons, and calculators, have also become smaller and lighter. Aluminum cans today weigh 30 percent less than they did two decades ago.

---

*Most work done on natural polyesters after the 1970s oil crises was dropped when oil prices came back down.

A typical U.S. car decreased in weight by about four hundred pounds between 1978 and 1988.

Information technologies make it possible to finely calibrate the amount of materials needed in manufacturing or in some cases to replace them entirely. Having run out of space for new roads, highway engineers plan to unclog traffic arteries through technology rather than construction—creating "smart roads" by installing sensors connected to information centers that will sort and send data to dashboard computers. Not only in cars but also in factories, information technologies such as computerized monitoring and process controls can help reduce emissions by changing the resource mix to cut back inputs or achieve a cleaner burn.

Clearly, however, decreasing weight has not led inevitably to decreasing aggregate waste levels. Some products, such as cars and tires, have become more durable, and people are keeping them longer. But along with the fact that there are more people buying more things, because cheaper, lighter products are often inferior in quality, or less expensive, or more difficult to repair, they are thrown away and replaced more frequently. Americans' consumption of shoes, for example, rose from 3.9 pairs a year in 1970 to 4.6 pairs in 1985. American hospitals now rely on disposables for 60 percent of operating-room equipment and 50 percent of items used in patient care. Where high-tech products are concerned, the acceleration of technological innovation has not only shortened their lifecycles, but has also increased scarcities of replacement parts, making outdated generations of computers, printers, or fax machines more difficult to repair. Obsolescence for scientific and technical equipment arrives ever more swiftly, with average useful life spans diminishing from fifteen years in the 1970s to eight in the 1980s and five in the 1990s.

In addition, new technologies often have unpredictable or double-edged consequences. The paradoxical effect of electronic information technologies on paper consumption is well known. Although the computer was supposed to lead to the "paperless office," between 1959 and 1986 the consumption of printing and writing paper in the United States jumped from seven to twenty-two million tons, while between 1981 and 1984, the use of paper by U.S. businesses rose from 850 billion to 1.4 trillion pages. Although plastics have decreased the weight of packaging, they have also opened up new realms of "convenience"—and waste. Clear polyethylene wrappers for magazines, for example, are widely replacing paper, or an address label directly fixed to the magazine cover, even as concern grows about the proliferation of nonbiodegradable single-use items.*

The advent of highly touted new materials also represents a mixed blessing for the environment. Weaving or bonding layers of plastic, metal, and ceramic fibers into composites by fusing together the properties of each component has yielded a boutique of new materials designed to increase strength, flexibility, lightness, and resistance to corrosion. Long incorporated into top-of-the-line tennis rackets, golf clubs, and speedboats, and more recently used in expensive customized specialties like artificial limbs and aircraft wings, composites now also make up about 8 percent of today's automobiles. As they enter the mass market,† mixed materials seem likely to revolutionize the way we build houses and infrastructure—bridges, mass transit, etc. As currently designed, however, composite materials seriously inhibit recycling.

Consumer goods now in use exhibit the same trend toward increasingly complex combinations of materials. A recent Office of Technology Assessment (OTA) study schematically depicted the nine separate layers made up of five different materials comprising a 0.002-inch-thick snack-chip bag. The study also noted

---

*In much of the industrialized world, the quest for convenience is also spurring energy-profligate air travel. Although flying consumes 40 percent more fuel per person than traversing the same distance by car, twice as much as by train, and five times as much as by bus, the airlines' share in U.S. travel grew from 4 percent of all passenger miles in 1950 to 18 percent in 1988. In 1990, air travelers covered 1.1 trillion miles, a 6 percent increase over 1989.

Other significant examples demonstrate increases in materials consumption resulting from institutional changes: When AT&T was forced to divest the Bell Telephone System, the recycling of virtually all American telephones ceased. When catalytic converters became mandatory in U.S. cars, the proportion of platinum recycled decreased sharply from the previous level: Whereas the jewelry industry (formerly the main consumer) had recovered almost all of the metal in use, no infrastructure existed to recover the platinum (now 34 percent of consumption) from discarded converters.

†American companies have led in developing composites, but are now investing less in the new materials than European and particularly Japanese competitors, who are gaining fast. Making the new materials competitive in price depends on devising process technologies, at which the Japanese have excelled.

the vast array of materials built into an automobile, including high-strength steel, aluminum, copper, ceramics, metal-matrix composites, and more than twenty different types of plastics.

The trend toward complexity in materials correspondingly increases the complexity of environmental trade-offs. Multilayer packaging prevents food waste and increases shelf life but increases nonbiodegradable trash. More durable components should make cars last longer but may also make repair more difficult. Although lighter cars and potato chip bags save energy, complex combinations of materials make both cars and packages much harder to recycle.

Whether benign or malign in environmental terms, these trends are unfolding independently of considerations about waste and pollution. The evolution of automobile technology clearly exemplifies the prevailing duality whereby we separate issues pertaining to economic productivity from those surrounding maintenance of the environment. Despite our recent concern about recycling, the increasing plastic component in cars already threatens the potential to recover materials, as scrap dealers see the proportion of profitable metals in junked vehicles diminishing and the amount of "fluff," including rubber, glass, and fiber as well as plastic, they must send to the landfill rising.

Recent articles on composite materials in *The New York Times* and *The Economist* failed to mention the inhibitions on recycling built into these new materials. Waxing euphoric about humankind's transition from being mere "shapers" of iron, stone, and wood to a new stage in which "you can design materials just as you can design the things you are going to build with them," *The Economist* rightly noted that fabricating composites requires integrated design of materials, components, and production. Missing entirely from the writer's picture of a future built on composites, however, was the notion of integrated design to avoid pollution and facilitate reuse of materials.

ᔦ

Promising new efforts to line up technical advances with the environment through design still have a livelier existence in concept than in reality. As an idea, "design for the environment" or "green design"—that is, the design of products and processes to minimize the waste of materials, eliminate toxic residues, and facilitate reuse and recycling—probably originated as recently as 1988, when the term reportedly was first used at a conference in Tokyo. Although currently a hot topic among in-house environmental specialists in some major U.S. corporations, in this country its effect on production processes, or strategies, is minimal. Few U.S. industrial enterprises appear to understand the linkage that the 3M Corporation summed up in the mid-1970s with the slogan "Pollution prevention pays." Most corporations spend far more time resisting environmental regulation than contemplating the potential efficiency of "green design." That won't change without a fundamental shift in macroeconomic policy.

If we can encourage the spread of integrated environmental design through market incentives, the redesign of products and processes is potentially the key to reducing waste. This approach represents the incorporation into business planning of various conservation and materials management options and goals. Although government programs aimed at pollution prevention and waste minimization are still largely hortatory, federal legislation in 1986 requiring reporting on toxic materials use and emissions, the Toxic Release Inventory (or TRI), has spurred a few forward-looking businesses to get ahead of the game.* Those instituting waste reduction programs include 3M, Dow Chemical (in 1986), Digital Equipment Corporation and Polaroid (1988), and Westinghouse (1989).

The pioneers see their efforts within an even larger context. Corporations like AT&T, Dow, Du Pont, and Hewlett-Packard are exploring a systems-oriented approach variously known as "industrial metabolism" or "industrial ecology," which views human economic activity as a system for the transformation of resources within the larger ecology of the biosphere. Proponents of industrial ecology argue that the unacceptably high resource inputs required by the industrial economy, including sinks for waste in the oceans, land, and atmosphere, make the cycling of materials imperative.

Advocates emphasize that "wastes" are raw materials. Thus, systems and products have to be designed from the outset to

---

*Some pressure was also felt after passage of the Hazardous and Solid Waste Amendments to RCRA in 1984, regulating hazardous waste.

reuse materials. Robert U. Ayres, a leading exponent of this approach, notes that "the history of the chemical industry is one of finding new uses for what were formerly waste products." To cite a prime example, it was the exhaustive search by German chemists for ways to use coal-tar residues generated by the gasworks of the early nineteenth century that led to the creation of the modern organic chemical industry.*

In a paper published by Arthur D. Little, management consultant Hardin Tibbs describes a remarkable example of industrial reuse of wastes in a town near Copenhagen, Denmark. Starting in the early 1980s, largely for economic reasons but later spurred by government-mandated cleanliness levels, companies in the town of Kalundborg set up an elaborate system of industrial environmental cooperation. Trade in wastes includes the exchange of steam and purified wastewater among an electric power–generating plant, an oil refinery, and a biotechnology production plant; the use of surplus gas from the refinery by a wallboard producer and the power plant; the use of sulfur removed from refinery gas by a sulfuric acid producer; the use of surplus heat from the utility by the town district-heating scheme and a seawater fish farm; the use by local farmers of sludge from the fish farm and from fermentation operations at the pharmaceutical plant; and the use of calcium sulfate derived from the power plant's sulfur emissions by the wallboard plant, as a substitute for gypsum!

With its network of cooperative exchange, Kalundborg undoubtedly sounds as far from achievable reality as Brigadoon to most American industrialists. Those currently essaying environmental design are focusing on immediate and more modest opportunities for waste prevention and materials management within their own plants. At the same time, they look anew at the bottom line, counting costs formerly ignored and overhauling accounting systems that had formerly buried energy, remediation, and disposal costs in general operating costs.

Businesses that pursue green design can employ a broad arsenal of methods to achieve their objectives, including reducing use of toxics, energy, and materials generally; incorporating

---

*Synthetic aniline dyes, phenolic resins, aspirin, and the sulfa drugs were all derived from coal tar.

recycled materials and components; designing more durable products; and designing for recycling, disassembly, and remanufacture. For example, considering what substitutes are available, using exact amounts of a given toxic in a manufacturing process may work better than eliminating that substance altogether. A process generating a large amount of waste that can be used elsewhere may be preferable to one that produces smaller quantities for which there is no use.

With green design, the interplay between technical achievement and consumer demand should be circular and continuous. Apart from cost savings, the environmental innovation often provides a decided marketing edge with environmentally conscious buyers. In addition it may confer a regulatory boon, which would itself increase sales. AT&T environmental engineer Braden Allenby envisions the advantage that might be seized by a hypothetical competitor who designed a widget that contained no lead solder (where his own widget used a lot) and was completely recyclable. Not only would the new widget be attractive to consumers, but it might spur a consumer trade-in program, which would severely disadvantage his widget if it could not be recycled.

Success stories may well entail financial gains as well as pollution and waste reduction. To recount a few achievements reported by U.S. corporations: One Westinghouse plant in North Carolina that processes stainless steel saved $290,000 a year through a 90 percent reduction in waste coolant in 1988 and another $67,000 through a subsequent reduction. Du Pont's Antioch (Ohio) Works reduced its hazardous waste by 95 percent between 1985 and 1991, thereby saving $5.8 million annually. Between 1983 and 1988, Amoco reduced its hazardous waste by 86 percent, saving about $50 million. 3M's savings on avoided pollution totaled $500 million between 1975 and 1989.

Outside the chemical industry,* a large part of the toxic waste

---

*Dow Chemical's energy conservation manager has comprehensively outlined techniques for reducing waste and saving energy that are specifically applicable to the chemical industry. These include reducing raw materials by improving quality of feeds, using inhibitors to prevent unwanted side effects, and using off-spec material where viable; maintaining the quality of mixing in reactors by distributing feeds evenly, premixing feeds before they enter the reactor, and improving mixing in the reactor; improving catalysts and providing separate reactors for recycle streams; minimizing the formation of waste prod-

generated by production processes stems from cleaning operations; thus, many of the possible reductions involve reducing cleaning steps and replacing chlorinated solvents with water or air, abrasive materials, aqueous detergents, alkaline solutions, acids, or biodegradable solvents like terpenes made from citrus plants and pine trees. Washington Scientific Industries, a precision machining company in Long Lake, Minnesota, was able to cut back the use of 1,1,1 trichlorethane (TCA) in the company's degreasing operations by 70 percent through installation of a high-pressure spray washer using a mild alkaline detergent, which was rapidly paid for by resulting cost savings. Similarly, Pacific Gas & Electric Co. reduced its hazardous waste from forty-eight drums in 1989 to twelve drums in 1990, in part by eliminating a parts cleaner in favor of a new cleaning unit using hot water and biodegradable soap.

A number of other successful cleaning modifications are relatively simple and low-tech. When solvent residues in its press wipers became a problem for its industrial laundry, the John Roberts printing company in Minneapolis, Minnesota, cut its solvent use by almost 75 percent by changing to a slower-evaporating wash for its presses and installing a centrifuge to "wring out" the wipers before they were sent to the laundry—in the process recovering about $34,000 worth of solvent annually. In other instances, intensive exploration of alternatives can lead progressively from reduction to elimination of toxics. After investigating ways to curb loss and evaporation of the solvent CFC-113 in cleaning circuit boards, the Canadian utility Northern Telecom developed a "no-clean" system, which instituted tight controls on materials sprayed onto the boards, thereby abolishing the cleaning step entirely, while at the same time improving soldering.

---

ucts in heat exchangers by cooling superheated plant steam, using lower-pressure steam, using on-line cleaning devices such as recirculated sponge balls or reversing brushes, reducing tube wall temperatures by installing a thermocompressor, and using staged heating to lessen degradation of fluids; installing sealless pumps; replacing furnace coils or replacing the furnace itself with an intermediate exchanger or by using existing superheat; improving separation in distillation columns; checking and modifying pipe sizes and composition; and using computer control systems to improve on-line control of flows, temperatures, pressures, etc., to analyze the process continually, to automate start-ups and shutdowns, and to program anticipated responses to upsets.

〜

Getting industries to design for recycling and reuse as well as conservation of materials is essential to achieving waste reduction. Here Germany is leading the way. The pathbreaking German environmental strategy to lay responsibility for waste directly at the door of industry was spurred by efforts to cope with mounting packaging waste. Packaging poses central technical as well as political dilemmas regarding incentives for manufacturers to reduce volumes of material that will end up as waste. Approaches to reduction within the industrial cycle include trimming product wrappings before distribution, as in the Digital Corporation's recent redesign of packaging for a "mouse" and for computer software that slashed volume by 89 percent and 88.2 percent, respectively. Upping quotients of recycled content in packaging also conserves materials.

When it comes to recycling postconsumer packaging, however, businesses have few incentives to reduce waste. They can affect their customers' disposal of packaging by designing for recycling: We have seen, for example, how far the recycling of plastics depends on the separation and labeling of resins. Yet, because industry does not have to pay for consumer waste disposal, and because the growth of the packaging industry has always depended on increasing waste, the current system offers few compelling economic reasons to redesign packaging.

Responding to that reality, in 1991 the German government passed what to Americans must seem a genuinely radical measure aimed at forcing industry to take responsibility for the collection and recycling of their products' packaging. As in the United States and other industrialized nations, postconsumer packaging constitutes about 30 percent of Germany's household waste by weight, and 50 percent by volume. Thus, the German law mandating the recycling of packaging aims to reduce the total residential waste stream by 20 percent to 25 percent. The ordinance covers all packaging materials, including the bulk of nonpaper consumer recyclables, including bottles, cans, and plastic containers and the plastic as well as paper wrapping that has heretofore been consigned to landfills or incinerators, as well as transport containers.

The law mandates collecting and recycling, in stages, first, of transport containers, second, of display packaging, and third, of containers and wrappings in contact with the product itself (sales packaging). The intent was to meet recycling targets ranging from 6 to 42 percent of most materials by January 1993 and 64 to 72 percent by July 1995. Manufacturers and distributors must take back transport packaging directly. (Stores have to remove display packaging prior to sale or provide containers for customers to dispose of it.) For sales (product) packaging, most industries and retailers have joined in an alternative collection system, for which they pay a fee: Their packages display a green dot, and these items are collected separately from households or at collection centers and processed for recycling by industry.

Because the recycling targets far exceed current German levels, or any other national projections, the feasibility of the program remains in question. In 1993, its first year, the dual system of collection and recycling ran heavily into debt. Moreover, German recyclables glutted European markets and quantities of German plastics were dumped in Europe and the Third World.

Critics also argue that by absolving consumers of direct responsibility the law retards changes in patterns of demand. Putting the onus for recycling on industry, however, appears the most direct way to use market pressures to internalize costs of disposal and thereby increase incentives both for reducing packaging and for making recycling cost-effective. Thus reduction, reuse, and recycling move upstream in product design and business planning.

At the start of 1994, the German government was planning to extend the target deadlines and investing heavily in processes for chemical recycling of plastics into fuel. Further ahead, the German Environment Ministry promises take-back legislation in a number of product areas in the next few years.* In response to what they consider inevitable longer-term imperatives, German auto manufacturers, including BMW and Volkswagen,

---

*The strategy seems best suited to higher-valued goods. Designing for disassembly of appliances, electronics, or automobiles has much in common with design for repair or remanufacture—that is, parts or components should be engineered so they can easily be detached from each other rather than being soldered together. They should also be refurbishable, rather than buried in a larger "throwaway" assembly.

are working hard now on designing for disassembly. Research is aimed at devising "snap technologies," uniform screws, sockets, and other built-in connections that can substitute for permanent bonds like welds, adhesives, or rivets; as well as systems for segregating and identifying the growing volume of plastics. Both German manufacturers have established pilot plants to explore disassembly and recycling, and BMW has declared its intention to produce an automobile built of 100 percent reusable/recyclable parts by the year 2000. In 1991, BMW introduced a two-seat roadster whose plastic panels were designed for disassembly and labeled with resin types to facilitate recycling.

Though some Japanese and American firms are following these initiatives, they are doing so with considerably less urgency. Japan's Nissan Motors and a consortium of the major U.S. auto manufacturers each has announced plans for research on facilitating plastics recycling and increasing the proportion of recovered materials in parts as well as auto bodies. In the United States, where the economics of auto recycling currently is based on the efficiency of powerful automated shredders, disassembly would represent a dramatic shift. Taking apart a vehicle by hand might take five person-hours. Given the growing plastic component in U.S. cars, and in view of the current rudimentary technical and market structures for plastics recycling, getting the labor-intensive process of disassembly off the ground in this country is likely to be particularly difficult. It probably will not happen without major new economic or regulatory incentives.

As we have seen, broader current trends in materials development are ambiguous at best, bringing both "dematerialization" and complex composites that resist reuse and recycling. In product development as well, short-term trends for the most part favor product differentiation and specialization, which increase the costs of repair and recycling. Meanwhile, most of the promising technologies for waste reduction linger on the drawing board, awaiting the market.

By itself, technology is not going to resolve the conflict between economy and environment. As a society, we must make choices. How will we sort it all out?

9

Can Science Sort It Out?

... certainty in the complexities of environmental medicine may be achievable only after the fact. ...

—U.S. Appellate Judge J. Skelley Wright, *Ethyl Corporation* v. *Environmental Protection Agency*,
March 19, 1976

As much as we might prefer to substitute "good science" for good policy, it just won't work. Over the past few years, the amount of human ingenuity devoted to studying waste and pollution has been prodigious. The further it takes us, however, the more clearly we can see how far we still have to go.

We have recently witnessed an explosion of knowledge, including the increasing application of cost-benefit and risk analysis as well as the new discipline of life-cycle assessment. These tools supply devices through which science and policy meet and they support environmentally sensitive choices in important ways. As we seek to determine the best public policy, these assessments can help in evaluating priorities and determining the economic and political costs of trade-offs. Still, they are often more effective at illuminating complexities than at helping to resolve them.

The real questions we face—about how much we should invest in preventing risks, given our vast ignorance of their causes and effects—must ultimately be answered through political decisions. Of course scientific analysis should support these choices as far as possible. In view of the traditional American impulse to resort to "science" to resolve conflict, however, it is worth making an effort to understand what cost-benefit, life-cycle, and risk analysis can and cannot accomplish.

The most commonly used measure, cost-benefit analysis, has serious flaws as an aid to environmental choices because it operates within a context that does not include many environmental costs and benefits. Assigning monetary values to the air, or water, or to species and their habitats, or to the quality of our own physical and spiritual life within this biosphere is difficult if not impossible. In addition, assessments of what natural resources are worth are stymied by our current system of accounting, which assumes that the future value of any object will be less than its current value. Even in nonmonetary, environmental terms the data necessary for establishing the benefits of pollution control often do not exist. A recent EPA report on costs of environmental controls explains why the agency did not attempt to compare costs with benefits:

> In the absence of controls, increasing population and levels of economic activity would have resulted in steadily decreasing environmental quality over time. In order to show environmental quality improvements resulting from pollution controls adequately, one would need to compare current levels of environmental quality indicators with estimated levels that would have prevailed in the absence of cumulative pollution control efforts. Except in the case of the criteria air pollutant emissions, such comparisons are precluded by the absence of data.

Moreover, the costs of pollution control are much easier to quantify than the benefits. We can compute fairly accurately what we spend on pollution abatement equipment, paperwork, and litigation. Calculations of productivity lost as a result of pollution control, though murky, are also generally accepted. The benefits of pollution abatement, however, are usually qualitative.

Even the costs are frequently overstated. Industry estimates often exaggerate future costs. Instances are not hard to find. In 1971 the oil industry calculated that phasing out lead from gasoline would cost $7 billion a year until the job was done. Actual costs ran about $150 million to $500 million a year—less than 7 percent of the industry estimate. A study in the mid-1970s by the Rand Corporation on the cost of reducing chlorofluorocarbons (CFCs) concluded the cost was infinite, that it could not be done at any price. But by the time an international treaty had been signed, in 1987, to phase out CFCs, the cost had come down to $50 billion, and in 1990 the EPA estimated the cost would be zero. Now it appears that the reductions will actually save money.

Life-cycle analysis (LCA) offers a valuable new tool for understanding impacts of products, processes, and activities on the environment.* In focusing on all impacts of the production and use cycle, such as extraction and use of materials at the front end and transportation, as well as of waste and disposal costs, LCA has increased awareness of impacts that are indirect, or that are merely shifted from one stage of the life cycle to another, or from land to air to water (thus the term "cradle-to-grave.") Where life-cycle analysis is used to target ways of using resources more efficiently, it has achieved demonstrable success, and the approach has become essential for industries seeking to understand environmental trade-offs as they endeavor to conserve materials and curb toxic emissions. LCA has proved more helpful for industry, however, where both objectives and data are rela-

---

*A Society of Environmental Toxicology and Chemistry (SETAC) report on life-cycle assessments describes the evolution of LCA studies. Life-cycle assessment has grown out of the "fuel cycle" studies of the late 1960s and early 1970s, which assessed the costs and environmental implications of alternative energy sources. From 1973 to 1975, oil crises led both the U.S. and British governments to commission energy analyses of industrial systems that would provide additional information on raw material requirements and solid waste emissions. Pathbreaking studies (called "resource and environmental profiles") were performed by the Midwest Research Institute (Franklin Associates) in Prairie Village, Kansas, and by Arthur D. Little, of Boston, in the mid-1970s. Although the concern about energy supplies faded along with the oil crises, interest revived in Europe in the 1980s, with the advent of the Greens and the efforts of the EC Environment Directorate to measure energy, raw materials, and waste involved in the use and disposal of liquid-food (i.e., drink) containers. As solid waste has become a political issue everywhere, the pace of life-cycle research has quickened, with studies routinely measuring gaseous, solid, and liquid emissions as well as energy, raw materials, and solid waste.

tively clearly defined, than it has in weighing questions bearing on public policy.

In view of the complex interrelationships, getting a grip on all the relevant data seems a well-nigh Herculean task.* Proponents, who are presently trying to standardize analytical methods, admit that current studies are often based on ill-defined assumptions and faulty data. In the public policy realm, methodological ambiguities are compounded by politics. Because "life cycles" include so many factors, interest groups can choose the ones that bolster their own case. Comparisons of products, as between disposable and cloth diapers, plastic and paper shopping bags, and metal and plastic plumbing pipes generally appear to find in favor of whoever pays for the study.

The diaper wars offer a prime example of the difficulty of reaching "objective" conclusions through life-cycle analyses. In the duel between manufacturers of disposable diapers, like Proctor & Gamble, and purveyors of cloth diapers led by the National Association of Diaper Service Industries, successive studies have accorded environmental superiority first to the disposables, then to the cloth diapers, again to disposables, and so on. When a number of consulting firms, including Franklin Associates and Arthur D. Little, compared raw material, energy, and water consumption; air emissions; water pollution; and solid waste, their conclusions about environmental impacts eventually pointed to a sort of draw, with cloth way ahead on raw materials and solid waste, and disposables solidly leading on energy and water consumption as well as pollution. Now that the furor has died down, the more relevant life-cycle question may indeed be: Should disposable diapers be recycled or composted?†

---

*According to the Society of Environmental Toxicology and Chemistry, LCA methodologies have focused so far almost exclusively on the "inventory" component ("quantifying energy and raw material requirements, air emissions, waterborne effluents, solid waste, and other environmental releases"), deferring work on "impact" and "improvement" analysis. Yet even requirements for the inventory seem dauntingly vast, including raw materials acquisition; manufacturing, processing, and formulation; distribution and transportation; use/reuse/maintenance; recycling; and waste management. Boundaries must be set. How far across the industrial structure or how far back in time should we go in summing up impacts from inputs used to produce more proximate inputs? And according to whose assumptions about current and future, positive and negative, values of products and pollution?

†Other prominent life-cycle cases include single-use versus refillable bottles,

The efforts of McDonald's to get in step environmentally present a different kind of "cradle-to-grave" ambiguity. Prodded by environmentalists, particularly the Environmental Defense Fund, about its expanded-polystyrene (Styrofoam) takeout clamshells, America's premier fast-food chain first tested the waters. After a trial program and preliminary planning with existing recyclers had fostered expectations that McDonald's would undertake polystyrene recycling, in the fall of 1990 the corporation reportedly acceded to pressure by the EDF and announced a switch to layered wrappers containing paper, tissue and polyethylene. Yet environmental conundrums continue. Because the layered wrap is not a good candidate for recycling, the switch appears less than a total victory for the environment. It does, however, represent a major reduction in the volume of waste and use of resources compared with the clamshell.

An article in *Science* magazine several months after the decision stirred the controversy further. Comparing paper and polystyrene cups, its author argued that paper uses more resources and emits more pollutants than polystyrene. The author argued that manufacturing a paper cup uses many more chemicals, equal amounts of hydrocarbons, twice as much water, and 36 times as much energy as a polystyrene cup while producing 580 times the amount of wastewater and 60 percent more air pollution. According to the EDF, however, a study done by Franklin Associates found that "the new wraps offer significant reductions in energy

---

and paper versus plastic bags. On bottles, the evidence appears to indicate a significant cost and environmental advantage for refillables. A Franklin Associates study showed that at current recycling rates among a variety of plastic and metal alternatives, a 16-ounce glass bottle refilled eight times generates the least air or water pollution; even if PET bottles were made of 100-percent recycled materials, if glass refillables were used 20 times, they would still have the lowest environmental impact. A study by the United States Public Research group also indicated that for most brands, beer and soft drinks are 20 to 40 percent cheaper when purchased in a refillable bottle than in a single-use container.

On plastic and paper grocery bags, an interesting comparison made by the Winnipeg Packaging Project finessed the direct comparison, finding that neither single-use plastic nor paper grocery bags fare well regarding energy use, even when high levels of recycling were achieved. A well-known Dutch government study, done for the plastics industry, comparing plastic and ceramic coffee cups also found in favor of plastic, largely because of the energy use as multitrip plastic bags in washing and sterilization of ceramic containers necessitated by public health and safety.

use and pollutant releases relative to polystyrene foam packaging.''

As we learn more about how to use life-cycle analysis, some of the uncertainties arising from questions of fact as well as from interpretations of value should diminish. For the foreseeable future, however, both the data requirements and differing methods for assessing and using information seem likely to frustrate precise conclusions.

Risk assessment poses different questions but involves similar dilemmas about uncertainty and value. The analytical procedure of risk assessment includes four elements:

- *hazard identification,* or data gathering
- *dose-response assessment,* or estimation of amount of exposure in relation to degree of toxicity
- *exposure assessment,* or estimation of actually occurring exposure of people
- *risk characterization,* or estimation (combining all previous elements) of overall health risk—for example, the added risk to one person in a thousand over a lifetime.

No one is very pleased with current efforts to calculate the risks to human and animal health and habitats. Like cost-benefit and life-cycle analysis, risk assessments are often shaped far more by the assumptions of their authors than by objective facts. Although conflicts about risk usually take the form of scientific disagreements, the pervading problem is scientific uncertainty. While a great many economic and social decisions ride on findings about risk, there is so much we just don't know. Moreover, once the scientists have had their say, then society must decide how to manage the risks that have been defined, weighing costs, legal requirements, politics, and public sentiments. Clearly, questions of relative value inherent in the "risk-management" process greatly increase the ambiguities.

Industry spokespeople and some scientists complain about the strict linear measures that commonly have been used in risk assessment. The most widely used method of gauging chemical risk, the "linear, no-threshold" model, looks at damage in rats and mice from a large, quick dose of a chemical, and extrapolates what that might mean to a human who gets a small dose over a long period. These analyses purport to gauge potential thresholds for adverse effects on human health—principally cancer—of various substances by looking at the animal studies of extreme short-term exposure to these substances. Critics assert that these assessments rely on worst-case scenarios, raising needless alarm among the public and leading to unreasonably conservative standards. According to toxicologist Alice Ottoboni, too many people appear to believe that "a chemical harmful in any amount is harmful in *every* amount." What the public really wants, experts argue, is impossible—a zero threshold of risk—and the economic costs of unrealistic standards are high.

Industry representatives also contend that stringent risk criteria can adversely affect the environment. Rigid standards about materials defined as containing toxics, they argue, may impede their beneficial reuse. Hazardous-waste regulations may cause industries to landfill or incinerate rather than recycle by-products that could be productively utilized.

Sewage sludge provides a case in point. Although the EPA supports the beneficial use of sludge, regulations promulgated in 1989 would in many cases preclude the use of sludge for soil enhancement. A scientific peer-review group created by the agency, including environmental organizations, academics, and local government officials, argued that the EPA's method of gauging pollutants from the perspective of a " 'Most Exposed Individual' . . . grossly exaggerated" risks. While the regulations would undoubtedly impede beneficial uses, there seemed little evidence that they would increase protection for individuals or the environment. The group summed up some of the frustrations endemic in conservative risk assessment:

> It is certainly easier to develop a scenario so bizarre that no person will ever have that much exposure and then to protect that nonexistent person and thereby all existing persons. . . . That approach leads to results that can be unrealistic, but where the degree of unreality cannot even be estimated.

A law passed in 1958 to ensure the safety of food supplies has aroused equally vexing questions. The regulation, commonly known as "the Delaney clause," bans any additive in food, including concentrations of pesticide residues, that has been shown to cause cancer in laboratory animals. The Delaney clause has not been fully enforced by the EPA for pesticides already approved and in use, in circumstances when the agency does not consider these substances a threat, but the regulation has stymied the approval of new, less toxic chemicals that do not pass the test. The law's problems are compounded by the fact that it directly conflicts with another regulation governing the use of pesticides on farms. Giving strong weight to the economic benefits of pesticides, the second law allows trace amounts of pesticides in food even if the compound has caused cancer in laboratory animals, as long as the residues do not pose an "unreasonable risk" to consumers. The existence of the two conflicting regulations, both embodied in law, has significantly impeded progress in setting useful standards. In her first official act as head of the EPA, in March 1993 Carol Browner asked for an updating of federal law to safeguard health without banning pesticides that do not pose risks, in effect loosening restrictions on pestisides in processed food and tightening them in regard to fresh produce.

Scientists (and the EPA) also assert that public perceptions and fears are often inconsistent with data establishing real risks. During the two decades or so that we have been using risk assessment, public fears about the hazards of industrial chemicals, particularly in causing cancer, have escalated. Between 1981 and 1988 the number of people who felt that groundwater pollution by hazardous waste or other chemicals was a serious problem grew from 28 percent to 54 percent. Because risk assessment has been used most commonly to compare different threats to human health, the environmental risks—to the biosphere, to animal and plant species and habitats, to the economic system, or to the quality of human life—are frequently ignored.

In the view of many scientists, this public concern with chemicals has drawn attention away from other environmental risks that may be more serious. These include destruction of habitats through the degradation of wetlands and soil erosion; loss of biological diversity; ozone depletion; and global climate change.

While water pollution is usually remediable, though at significant cost, the decline of species and destruction of habitats, including our own atmosphere, may be irreversible.

Further, the priority we have given over the last decade to cleaning up hazardous waste has rarely flowed from judgments about actual health or environmental benefits. The EPA estimates that about 1,000 cancer cases each year result from exposure to toxic waste sites. The $1.75 billion spent annually to clean up Superfund sites amounts to about $1.75 million a case. By contrast, the breast cancer research program of the National Cancer Institute spends what amounts to about $760 a case (given around 176,000 breast cancer cases a year). Spending on Superfund also dwarfs expenditures for programs crucial to maintaining species and habitats, like wetlands protection, or those likely to pay major future dividends, like pollution prevention.

Environmentalists, on the other hand, argue that the "unreasonable risk" standard, employed by the EPA since the passage of the 1976 Toxic Substances Control Act (TSCA), sets impossibly high hurdles in determining whether to regulate existing risks. The standard views regulations in relation to their economic costs and mandates the EPA to find the least burdensome way of regulating any problem substance. In effect, the measure requires the agency to "prove" that the harm from any given toxic substance is bad enough to justify the costs of required controls to business or the national economy.

Given the multitude of chemicals now in use, and with a thousand or more coming on the market every year, determining actual hazards is a daunting task. Although progress has been made in gauging the risks of certain chemicals in causing cancer, data on the toxicity of many pollutants is seriously deficient. In 1991, the General Accounting Office reported that 15 years after the enactment of TSCA, tests had been completed on only 22 chemicals and the EPA had assessed the results of only 13.*

Pesticides are some of our most widely used chemicals, yet

---

*Over 100,000 halogenated carbons (HHCs), persistent, bioaccumulative chemicals used in industrial processes and as pesticides, are registered in the OECD countries, but only 3,000 can be identified using current analytical methods. Nor do those registered include the 500,000 by-products resulting from reactions occurring when these chemicals are introduced into the environment.

according to the National Research Council (NRC), the health effects of most active ingredients in pesticides are not known. In 1984 data were available to assess the health effects of only 10 percent of the ingredients in pesticides. Most pesticides were registered before their potential for causing cancer, birth defects, and other chronic health problems had come under scrutiny, and few have been tested in light of contemporary knowledge. Currently, about 620 active ingredients and 23,000 pesticide products require reregistration. In 1993, only 29 pesticide products had been fully tested and reregistered. Perhaps the most crucial issue, the synergistic effects of various chemicals used in tandem, has not been well evaluated.

As the EPA has pointed out, professional judgments themselves are based on "incomplete and often inadequate knowledge about 1) the extent of human and ecological exposures to pollutants and 2) exposure-response relationships." Most studies of risk to human health have involved laboratory animals. Although, according to the National Research Council 30 percent of the insecticides, 50 percent of the herbicides, and 90 percent of the fungicides used on American farms do cause tumors in laboratory animals, available data do not show pesticide residues in the average diet as a major contributor to "the overall risk of cancer for humans." Human immune systems appear to be stronger and more resilient than those of rats and monkeys. Having said that, we are nonetheless left vastly ignorant about human responses to chemical impacts.

The controversy surrounding the aromatic organic chemical dioxin illustrates some of the problems with establishing the relationship between pollution and risk. Dioxin is well known as a by-product of incineration and also the main toxic ingredient in the defoliant massively applied in Vietnam, Agent Orange. Known to cause cancer in laboratory animals and often considered to be among the most toxic compounds known, the level at which dioxin harms human health has not been confirmed. Although the spraying of the chemical on roads around the town of Times Beach, Missouri, led to its inhabitants' permanent evacuation in 1983, today many scientists consider government reaction to the Times Beach contamination "alarmist."

In response to industry complaints,* the EPA is reassessing its standards, the world's strictest, for dioxin in drinking water and for average daily consumption. In an interim report issued in September 1992, a federally appointed panel of scientists found the risk to Americans from dioxin ingested in beef, dairy products, chicken, and fish lower than once thought, but the risk to chemical workers handling high doses greater than previously believed. One finding emerging much more clearly as a result of the panel's effort to look beyond cancer in people is the biological havoc caused among fish, birds, and other wild animals by even minute levels of the chemical.

Furthermore, as we have seen, we are only beginning to understand the quantitative extent of our chemical exposure. Over 20 separate national surveys, each established for a different purpose, collect data on industrial waste streams for different waste generators, waste types, and time periods. For example, the most politically significant database, the Toxic Release Inventory (TRI), requires manufacturers in 20 industries to report annually to EPA on their releases of 322 chemicals or chemical categories. Hundreds of chemicals listed as toxic under other environmental laws, however, are excluded from the TRI report.

Moreover, the index captures only the contaminants emitted or discarded by industry—not the much larger quantity embedded in products that are thrown out by consumers, or stemming from nonmanufacturing sources such as mining, waste treatment, farming, and government facilities. Including consumer products as well changes the picture greatly: The New Jersey Department of Environmental Protection and Energy reported that in 1990 at least 83 percent of the cadmium, 92 percent of the nickel, and 99 percent of the mercury used by industry went into products that for the most part would later be disposed of as wastes, none of which was counted among the heavy metals emissions reported by businesses.

Another source of uncertainty is the lack of knowledge about exposure to toxics over time. The commonly used "linear, no-

---

*Dioxin-containing products are no longer manufactured in the United States, but the chemical is still emitted and discharged into surface waters as a by-product of the chlorine bleaching of pulp and paper products.

threshold" model for risk analysis focuses on large doses of one chemical, swiftly administered. Few studies have adequately evaluated chronic exposure; to do this, the experimental design must have a far longer time frame. Yet the effects of air or water pollution on the health of humans and other species and habitats may only become apparent after many years. Thus our understanding of the individual effects of most chemicals flowing through our economy remains highly incomplete.

Nor, as we struggle to assess a few specific chemicals, have we begun to fathom the effects of exposure to several different toxics simultaneously. Evidence emerging now points to the subtly debilitating effects on animal life of low-level exposure to a spectrum of chemicals. Unlike the massive die-offs of birds as a result of DDT sprays in the 1950s and early 1960s, the abnormal development of fish and bird life in and around the Great Lakes and the St. Lawrence River appears to result from chronic exposure to three separate families of industrial emissions: pesticides, aromatic hydrocarbons, and heavy metals.

Moreover, chemical impacts on hormonal and immune systems have been assessed only minimally, compared with studies of cancer. Wasting diseases afflicting fish and wildlife around polluted urban bays suggest that low-level chemical exposure resulting in long-term accumulation in the body appears to have corrosive effects on the immune and reproductive systems. Though not as dramatic as the threat of cancer and mortality that have been the focus of public policy heretofore, these subtle assaults on bodily systems may represent a more pervasive risk.

More broadly, our track record with human-induced environmental risks generally must give us pause. Public insecurities get their impetus as much from a general sense of the historical record as from data on current conditions. Far too often experts have been surprised by and have resisted acknowledging, the advent of new environmental hazards resulting from technological innovations. Because industrial, military, and nuclear waste were so long largely out of sight and mind, successive revelations brought successive shocks.

Further, when we do try to estimate the effects of technologies and processes, population growth and material expansion soon make these calculations obsolete. More information often brings more pain. The loss of protective ozone as a result of chlorofluorocarbon emissions was not foreseen, and not widely perceived until the destruction was well advanced. In 1992, ozone loss over the Northern Hemisphere again was twice as great as scientists had estimated.

Awareness that many of our dilemmas about environmental risk stem from uncertainty doesn't relieve us of the necessity to make choices, both for the present and for the future. On the one hand we do not want to put our economic lives on hold until we take complete measure of all possible risks. Nor do we want to sacrifice innovative resource management to overly rigid regulation. On the other hand, we do not want to mortgage our future by ignoring threats not yet "scientifically verified."

Where human impacts on the environment are concerned, time appears to be a double-edged sword. It often takes quite a while for damage to become evident. With oil spills like that of the *Exxon Valdez*, and pesticide and chemical emissions like DDT and PCBs, we have to wait until the effects work their way through the fish-spawning and animal-breeding cycles before we can gauge the extent of the injury. By then, however, it may be too late to restore the health of the affected species.

Both the public and the experts need information to guide decision-making. Public questioning of experts has on balance proved useful, not only in slowing environmentally destructive "progress" but also in pushing scientific inquiry to follow changing values. A better-educated citizenry should be better able to distinguish between significant and trivial risks. Yet ultimately, given the interrelatedness of all physical things within the ecosystem, our questions turn out to be almost as complicated as the meaning of life itself. Rutgers ecologist David Ehrenfeld explains that the problem with relying on the physical sciences to answer our environmental dilemmas lies in the gap between the closed systems that work best for scientific modeling and open-ended human and environmental systems:

> Rarely is expert knowledge sufficient for analysis, prediction, and management of a given [environmental] situation. This is because in order to limit the number of variables they have to contend

with, experts make the assumption that the systems they are working with are closed. Real-life systems are hardly ever closed. . . .
The reductionist methods of science, which can work extremely
well in closed systems, tend to break down under the open-
endedness imposed by biological complexity and by the interact-
ing complexities of political, economic and social systems.

Similarly, as solid waste consultants Jeffery Morris and Diana
Canzoneri point out, "Determining complete lifecycle impacts
of a product requires bringing together in one place the same
web of complex information that is required to run a centrally
planned economy." We know how well that works. Thus, even if
we spent vastly increased amounts of our resources on cost-
benefit, life-cycle, and risk analysis, all the information we
amassed would still not settle these issues for us.

10

~~~~~~~~~

The Business of Business

*Technology and production can be great benefactors of men,
but they are mindless instruments.*

—CHARLES A. REICH, *The Greening of America*

DESPITE ITS WARTIME feats in meeting national challenges, we cannot rely on the private sector to save us this time. As much as our tradition predisposes us to rely on entrepreneurial ingenuity and energy to solve most problems, U.S. business clearly is not poised to wage a war on waste. The economic incentives driving marketplace decisions still heavily favor exploitation of our habitat and encourage resistance to pollution controls. Unless government takes a strong lead to shift the direction of incentives, market forces are more likely to savage the environment than to salvage it.

Of all the advanced industrialized nations, only in America is it really necessary to argue the need for government policy to reshape institutional responses. In both Europe and Japan, political culture governing business–government relations differs markedly from our own. For citizens of the other advanced industrial areas, the role of government in setting institutional directions is widely accepted.

In the volatile and interdependent world of the 1990s, political cultures may be changing—and converging. Yet even if that happens, the differences have been so thoroughly embedded that they will continue to shape our behavior for some time to come.

In Japan, society functions as a web of firmly interconnected institutions. From the top, government sets goals for business and actively intervenes to facilitate dealmaking and foster research. For that reason, when a government ministry decides to implement a new policy, the change takes place relatively quickly.

To take a case in point, in the early 1970s, heavy air pollution in Japan disappeared rapidly in the face of new environmental regulations. Although long overdue, once these laws were promulgated, they swiftly curbed emissions from vehicles and plants. The new rules required heavy investment by industry in smokestack scrubbers and water purification devices. Despite the costs, Japanese businesses undertook the changes energetically and then went on to dominate international markets in a number of pollution-control products. In the same way, after the oil shocks of the 1970s, Japan's powerful Ministry of Trade and Industry (MITI) mandated a high degree of energy efficiency. Subsequently, Japanese industry sharpened its competitive edge in energy technology.

In Germany, though efforts to integrate the reunited east and west now strain the country's political institutions, the German people's approach to governmental authority still appears destined to resemble that of the Japanese more than our own. To an American eye, boundaries between separate sectors—public and private, business and labor, markets and politics—in both countries blur confusingly. In Germany, business has a responsibility for social order, and both the government and the unions have a say about organizational decisions in the private sector. The three together, for example, manage the system of vocational education that produces the highly skilled German work force. The flexibility flowing from this broad, consultation process made possible Germany's swift implementation during the latter 1980s of the most ambitious environmental legislation any nation has yet attempted. As social dilemmas intensify, business-union confrontations will sharpen, but the emphasis on consensus will continue to underlie negotiations.

The Netherlands also relies heavily on a dialogue between business and government, usually preceding parliamentary action, to achieve voluntary agreements about regulation. In May 1989, after the country's government became the first to fall in a crisis over the environment, the Dutch system produced the most comprehensive environmental plan yet to emerge in any advanced industrialized country. The Netherlands National Environmental Policy Plan mandates industry to meet 60 percent recycling levels by the year 2000. Similarly, the French government is working with major industries to design strategies for reducing waste. Even in the United Kingdom, the birthplace of Lockean individualism, where adversarial relationships between public and private sectors most closely resemble our own, the broad authority of government is accepted quite matter-of-factly throughout British society, giving the highly respected British civil service significant leverage in resolving thorny issues with industrial leaders before they reach the stage of public debate.

After more than a decade of probusiness administrations, many Americans would be surprised to know to what extent the complexities of both our regulatory process and the standards themselves grow out of a tradition of hostility between business and government that is far greater here than in any of the other advanced industrial nations. One reason the relationship in Europe and Japan has seemed relatively harmonious (to our eyes) is that the citizens of the other advanced industrialized nations have been willing to entrust the management of regulation to civil servants, who generally enjoy a degree of esteem within their various societies that American government employees do not. Another is that in these societies open debate is to a much greater extent subordinated to discussions among the major players, in which citizens groups play a significantly smaller role. Everyone involved works at hammering out an agreed position, in effect writing the regulations before they are legislated or promulgated. Once those at the table reach a consensus, they are not expected to reopen the issue, but to make the rules work.

In this country, on the other hand, regulations evolve through a much more open, combative, and lengthy process. Government, business, labor, the scientific community, the environmental movement, and others in effect thrash out the issues at length in an open forum. Given the prevailing mistrust of government and the institutionalized adversarial relationship between government and business, all parties rely heavily on "expert" or "objective" analysis of risks and benefits to bolster their authority.

When federal agency officials write the regulations after public hearings, it is only the beginning of the struggle. Congress plays a major role in mandating the structure of the rules and the way they are supposed to be implemented, through complex rule-making procedures, economic analysis requirements, and implementation deadlines. Businesses lobby heavily all the way and, following the promulgation of the regulations, continue to contest them in the courts. To prevail in the American arena depends on investments in expertise, publicity, lobbying, and litigation, and decisions can be endlessly reopened where resources to do so are available.

The prospect that disputed issues will ultimately be resolved by a judge shapes the behavior of U.S. business throughout the regulatory process. Because their statements and actions are likely to become part of a court dossier, parties are deterred from cooperating earlier for fear that this may later be seen as validation of their adversaries' claims. Viewed as potential legal precedents, minor issues become major battlegrounds. When the rules themselves are but a way station to the courts it should not be surprising that regulatory hearings function chiefly as platforms for different parties to take public positions.

The regulation of the carcinogenic gas vinyl chloride (VC), used in making plastics, graphically illustrates the difference between the American style and those of its fellow industrialized nations. Industrial analyst Joseph Badaracco uses the vinyl chloride case to contrast the U.S. process with what he calls the "cooperative" procedures in Great Britain, Germany, and France. In the latter countries, business, government, and labor worked together to resolve the VC problem, on much the same terms as those ultimately reached in the United States. In the United States, on the other hand, industry fought the Occupational Safety and Health Administration (OSHA) unremittingly on the "zero-acceptable" limit that it was advocating.

The American tug-of-war began with B. F. Goodrich's announcement of suspicious employee cancer deaths possibly traceable to VC, and ended two years and many million decibels later.

In January 1974, B. F. Goodrich drew attention to three recent deaths from liver cancer among workers at its Louisville, Kentucky, VC plant. Over the following nine months, industry, labor, the scientific community, and OSHA all produced expert testimony and competing studies about health risks. Along the way, industry predicted that a shutdown of VC-PVC production would cost 1.6 million jobs and reduce the gross national product by at least $65 billion. After OSHA's June 1974 hearings and the submission of additional information by interested parties, the VC record exceeded four thousand pages and included over eight hundred oral and written submissions. OSHA published its final standard in October. Two minutes after the OSHA announcement, the Society for the Plastics Industry, Hooker, Union Carbide, Firestone, Tenneco, Air Products, and B. F. Goodrich—with lawyers standing by in telephone booths in Washington and at courthouses around the country—had all filed suits in various circuit courts.

Ultimately, after considerable corporate expenditure to resist regulation, compliance came relatively cheap. A year and a half after their rush to court, the plastics and rubber industries had found the costs of meeting the standards much less than they had previously estimated. By April 1976, the VC and PVC industries had reduced emissions almost to zero—at a cost estimated at only $200 million to $280 million, and a loss of 375 jobs.

In this country during the last decade, the antigovernment populism effectively orchestrated by President Ronald Reagan and later by Ross Perot heightened business resistance to government oversight. Both popular leaders played effectively on traditional American suspicions that government is more often the problem than the problem-solver. Even under an activist President such as Bill Clinton, efforts to steer society in new directions run smack into this widely shared assumption.

The corollary belief in the ability of the private sector to deliver societal needs most effectively fared less well during the 1980s. In that decade, the performance of U.S. business showed that loosening government oversight and increasing laissez-faire does not inevitably foster the greater economic good. While manufacturing growth did increase for a few years, responses on the part of corporations included intense efforts to buy up rivals coupled with slackening concern about development of new products and services; record corporate debt resulting from takeover raids and defenses against them; massive investment in commercial buildings and shopping centers, many of which still stand empty; the failure of vast numbers of deregulated S&L's whose officers had personally prospered from this construction; and soaring executive salaries concurrent with record layoffs of middle management and shop-floor workers.

In the current period, management strategies appear to be working toward strengthening the U.S. economic position. Nonetheless, change will come haltingly in the environmental area until incentives shift radically. At bottom, the responsibility of business is to generate profits, not to improve public welfare. Although conservation of materials and energy would seem to enhance profitability, the thrust of U.S. resource policy is to secure an abundant supply for cheap exploitation. That policy has shaped business strategies for the last hundred years.

Since the mid–nineteenth century, a variety of policies supporting mining, particularly of fossil fuels, and forestry have tilted the playing field decidedly against conservation and recycling. Much of our law that allocates resources was shaped in a time when Americans urgently sought to fill the land and exploit its bounty. Ever since the West was divided up into acre parcels, implicit and explicit government subsidies have urged Americans onward to develop the land, even seashores and floodplains. The 1993 Clinton-mandated hike in fees for grazing public lands to $4.28 an acre amounted to a rise from one fifth to a mere one third of the private market price. Under the General Mining Act of 1872, which until the spring of 1994 continued to regulate mining claims, those who find hard-rock minerals on public land can buy the land for virtually nothing (five dollars an acre or less). Extractive minerals-industry tax preferences are also still significant. "Depletion allowances," permitting deductions of part of the value of production, were extended as emergency measures to some minerals industries during World War I, to all primary minerals producers during the Depression, and to non-metallic minerals producers during World War II. Though the benefits of these allowances are substantially reduced by state taxes, the Office of Management and Budget estimated that the tax break amounted to $505 million in 1991. In 1993 the federal budget allotted $560 million in subsidies to the mining industry.

Similarly, timber producers reap government windfalls in below-cost purchase of timber stands. Taking into account U.S. Forestry Service expenditures on forest management, surveying, road building, and firefighting, estimated at $1.2 billion a year between 1982 and 1987, the U.S. Treasury has subsidized the exploitation of national timber by taking losses of $200 million to $500 million annually on timber sales. The forest products industry also receives favorable tax treatment for some costs of managing timberlands (amounting to an estimated $256 million in 1986) and for investment in reforestation (estimated at $203 million for 1989).

But the most important subsidy to primary materials production—and the principal barrier to change—comes from energy subsidies. All told, existing subsidies for energy, largely oil, coal, gas, and nuclear, were estimated at approximately $36 billion in 1989. Though the Energy Policy Act of 1992 contains positive incentives for renewables and efficiency, and some tax benefits that applied in 1989 are being phased out, the effective subsidy in 1993 was only slightly smaller. Even as their glory days may be ending, fossil fuels continue to enjoy the many benefits of long-entrenched power. Our prime energy user, the automobile, benefits from preferential treatment for gasoline, government-built roads, and tax breaks for free parking. Department of Energy research programs reserve a hefty share—$750 million, approximately 40 percent of research monies in 1991—for new techniques in fossil-fuel production and cleanup. (Around 30 percent went for research on nuclear power, while conservation and renewables shared the remaining 30 percent.) The oil industry also avoids substantial cleanup costs through its exemption from most federal environmental laws including Superfund. Despite oil companies' ferocious resistance to an energy tax in 1993, a rise of 4.28 cents a gallon was enacted—but U.S. gasoline taxes are still the lowest in the industrialized world and U.S. energy prices are 30 percent lower than those of a decade ago.

In agriculture we find an analogous situation, in which long-standing policies support the expansion of grain cultivation requiring heavy infusions of chemical fertilizers and pesticides.

Department of Agriculture commodity-price and income-support programs, which shape farming practices on more than two thirds of U.S. cropland, base support levels on previous production of heavy agrichemical-using crops: corn, wheat, cotton, and soybeans. Furthermore, the prevalence of crop specialization tends to increase pests and deplete soil quality, further heightening U.S. farmers' dependence on pesticides and fertilizer.

At the same time, the insistence of Department of Agriculture credit and insurance agencies on "accepted practices" makes it very difficult to use alternative methods like crop rotation to reduce erosion. Both the Farmers' Home Administration (FHA) and the Federal Crop Insurance Corporation (FCIC) condition farm loans and insurance on the use of "good farming practices"—that is, those associated with conventional, high-input farming methods. A farmer who had been growing corn over a number of years would stand to lose government supports, FHA credit, and FCIC insurance if he converted to a multicrop rotation including, say, corn, soybeans, alfalfa, and meadow grasses. Urged by an agricultural extension agent to try some new techniques, one farmer summed up the impasse when he replied, "I already know how to farm a lot better than I'm doing."

In the summer of 1993, the EPA announced an unprecedented effort at cooperation among the agency, the Department of Agriculture, and the Food and Drug Administration to curb the use of pesticides. Yet Agriculture's research budget continues to heavily favor traditional methods. In 1991 the DOA allocated only $6.3 million out of a $500 million research budget (about 1 percent) for alternative approaches. (On the other hand, the department spent $45 million that year promoting the sale of beef cattle, which are by far the heaviest consumers of grain.)

While all this support for resource exploration and chemical dependency continues, the incentives for pollution control are largely negative ones—avoided costs. Because the commitment to source reduction is largely hortatory, the connection between pollution and wasteful exploitation of materials is not clearly established. American businesses are getting decidedly mixed signals. Sanctions for disobeying rules, liability for past infractions, and the threat of public exposure pressure businesses to toe the line. Yet policies favoring continued exploration for cheap materials discourage them from pursuing the most effective way to curb pollution and waste—by reducing materials inputs.

Nor does business have significant incentives at present to develop pioneering environmental technologies. For several reasons, the private sector's role in leading-edge environmental research is likely to be limited. First of all, environmental transactions—such as the deposit of ash from incinerator combustion on the surrounding neighborhood—take place largely outside the market. Because environmental costs are unpriced and still largely unreimbursed, the profitability of environmentally superior technologies depends on relatively weak and unpredictable incentives in the form of legislated regulations. Beyond that, as we have seen, existing public policies in energy and agriculture stack the deck heavily against environmental innovation.

Second, as World Resources Institute economists Robert Repetto, George Heaton, and Rodney Sobin point out, before critical environmental innovations become possible, techniques and knowledge with wide applicability in a variety of industrial contexts must be developed. In many instances, these "generic technologies," like catalysts for neutralizing chlorine, or microbes for settling heavy metals out of sewage, become a kind of common good, preceding the emergence of products, processes, or systems that can be appropriated competitively. Because investors will not be able to capture these technologies directly, the market is unlikely to finance their development. The WRI economists note that generic technologies are vital for advances in pollution prevention through improved catalysis (allowing more benign feedstocks, or removal of toxic by-products); separation operations (distillation, drying, cleaning, degreasing, and evaporation involving vast quantities of solvents and other hazardous materials as well as the promise of improved pollutant filtration); precision fabrication; development of new materials; and monitoring and control mechanisms underlying "intelligent manufacturing systems."

Given these disincentives, incorporating conservation into business planning and product design often requires a major departure from established practice. From U.S. businesses' short-term perspective of annual balance sheets, the costs of retooling too often loom larger than the benefits of increased efficiency. In 1992 Ford Motor Company's Lorain, Ohio, plant still used twice as much energy to produce a vehicle as Japanese and German automakers. According to a Ford spokesman interviewed in *The Washington Post,* the company deferred acquisition of new energy-efficient motors that are used in many Japanese plants because the energy savings would not pay for the new equipment quickly enough. Ford also curtailed a program to install high-efficiency lighting throughout its mammoth plant, which is about 20 percent larger than a comparable Japanese plant, for the same reason.

Simple institutional inertia also impedes managers from changing the corporate directions on which they have staked their careers. From the 1960s through the 1980s, corporate culture within major U.S. enterprises discouraged deviation from goals set through established planning mechanisms, and it is not clear that the "leaner, meaner" 1990s will see widespread organizational innovation. EPA energy specialist John S. Hoffman argues, for example, that U.S. corporations, with their fondness for quantitative tools including mathematical "inventory theories," would have rejected out of hand the "just-in-time" inventory systems that proved so successful in Japan had they been suggested here before the Japanese tried them.

In the 1980s, U.S. businesses all too often sold off innovative environmental technologies, most frequently to Japanese and German companies, rather than invest the capital and effort to bring them to market. Ostensibly these sales were made to improve balance sheets, but they appear to reflect even more strongly the reluctance of managers entrenched in established organizations to shift goals and reorient careers behind new ventures.

Curtis Moore, former counsel to the Senate Committee on Environment and Public Works, has calculated some of the losses. Item: In 1980 the United States was the world's largest manufacturer of photovoltaic cells, which produce electricity from sunlight. Following ARCO's divestment of its leading-edge solar division in 1989, the U.S. has fallen behind Japan and Germany in this area. Item: U.S. power plants are now purchasing a new multibillion-dollar pollution control system from Japan and

Germany that was developed in the United States by Corning and sold to Mitsubishi, which licensed it to Siemens when Corning failed to find a U.S. market. Item: Fuel cell technologies, which greatly increase efficiency in generating electricity, originated in the U.S. space program, but their production is now monopolized by two Japanese firms, Fuji Electric and Toshiba.

For businesses to successfully incorporate green design into manufacturing processes, top leaders must be committed to change throughout the system. Bringing it off requires gaining access to new sorts of technical information, retraining engineers to include pollution and waste prevention in their original programs, and changing accounting systems to reflect costs of disposal. The people who make the capital investments in new systems are seldom the people who use the systems. In the calculations of decision-makers at the top, expenditures at the front end typically overshadow potential savings in running costs. Thus company leaders themselves have to be highly committed to making pollution prevention work throughout the organization. Unless incentives change, this sort of commitment will depend on—very likely rare—seizures of social concern.

For major corporations it must often seem easier to spend money lobbying to forestall political action than to invest capital in new technologies. It is well-nigh impossible to calculate accurately the corporate resources that are spent buying political influence, since much corporate money is carefully channeled through law firms and other agents outside actual donor firms. Yet the sums are clearly significant, accounting for a major share of campaign spending for contests costing, in 1992, on average, $550,000 for a House seat and more than $3.2 million for a Senate seat.

In an unusually detailed description of energy and auto industry lobbying in connection with the National Energy Security Act of 1991, the U.S. Public Interest Research Group (PIRG) catalogued the contributions of the oil, coal, nuclear, and auto-company PACs to the Senate Energy Committee members who shaped and passed the bill. The legislation called for opening up the Arctic National Wildlife Refuge to oil and gas exploration; restricting citizens' rights to participate in nuclear power plant licensing; deregulation of the electric utility industry; and the easing of Clean Air Act pollution standards for coal-fired plants.

Although the bill was ultimately "neutralized" on the Senate floor by the deletion of Arctic drilling provisions, the committee version clearly reflected the more than $3 million poured into the committee members' campaign coffers by the energy, nuclear, and motor vehicle industries.

According to the PIRG report, the PACs got what they paid for (and paid for what they got). Senators J. Bennett Johnston, Democrat of Louisiana, and Malcolm Wallop, Republican of Wyoming, who sponsored the bill, received $355,619 and $217,717, respectively, from energy- and auto-industry PACs between 1985 and 1991. Those same PACs contributed a total of $3 million to members of the committee during the same period. The payoff is clear in the record. An average of $69,511 went from nuclear-industry PACs to senators voting to restrict public hearings on nuclear reactors, while those voting to retain public participation got on average only $24,146. Senators voting to open the Arctic Refuge to oil exploration received an average of $61,033 from oil-industry PACs, while those voting to protect the refuge averaged only $17,226.

Inevitably, industries' intensive lobbying strategies cloud their perspective on their own future advantage. Just as we find it difficult to pat our heads and rub our stomachs simultaneously, the effort poured into resisting environmental legislation has weakened the pursuit of energy and resource efficiency by U.S. business. The auto industry, for example, continues to compete heavily at selling high-performance wagons and pickups in the suburbs while resisting higher standards for fuel efficiency mightily. Even for the designers of GE's prototype electric vehicle—developed in response to California's requirement that 2 percent of each car company's sales be "zero emission vehicles" by 1998—"performance" has translated into speed. Engineered to compensate for the perceived passivity of electric power by substituting "real muscle," the GE Impact beats most gas-powered autos in acceleration. Priced at $100,000, however, it is not likely to put much of a dent in California's smog problem.

American industry's declining investment in industrial research since the mid-1980s reflects the priority of current balance sheets over future growth. Growth in R&D investments by the private sector has fallen sharply as a result of the huge increase in corporate debt during the 1980s as well as the increasing weight of institutional investors demanding immediate returns. During the 1980s our chief competitors in Japan, Germany, France, Italy, Sweden, and Britain upped their private-sector R&D investment, cumulatively spending 34 percent more per capita than the United States by 1991.* Though its population is only half our size, and its GNP two-thirds as large as ours, Japan until 1992 invested in new factories, machinery, and other capital goods at nearly twice the per capita rate of the United States, and investment in R&D by private industries equaled ours even in absolute terms in the early 1990s. With the recent Japanese recession and European downturn, growth in private sector R&D tapered off. In 1993, however, the Japanese government signaled its continued commitment to research with a 6 percent hike in budgetary support for industrial R&D.

While U.S. business was selling off significant aspects of the national competitive edge in environmental technologies to European and Japanese concerns, U.S. industry was losing ground more generally in innovative environmental areas. Whereas the distinctly muted concern for the environment of the Reagan and Bush administrations set the tone for U.S. business, in Germany, society as a whole was shocked into a communal environmental zeal, first by the acid-rain-induced death of great stands of beloved national forest, noticed initially in 1979 or 1980, and again by the Chernobyl nuclear meltdown of 1986. At the end of the last decade, politicians in Germany's political mainstream co-opted Green Party policies for their own programs. In addition to its packaging and other "take-back" programs, the country instituted the most rigorous environmental controls of any nation in the world, including the United States. The Germans have retrofitted all their power plants to slash the "Nox and Sox" (nitrogen and sulfuric oxides) that are the chief cause of acid

---

*From the mid-1980s to the early 1990s U.S. industrial investment in R&D slowed from an average annual growth of 7.5 percent between 1980 to 1985, to only 0.4 percent between 1985 and 1991. From a peak in 1989 of $78.73 billion, it dropped to $77.84 billion in 1990, the largest one-year drop in three decades. The Europeans and Japanese spent 25 percent more per capita than the United States in 1988 and 34 percent more in 1991.

rain, while committing industry to a total ban on CFCs by 1995 and a 25 percent reduction of carbon dioxide emissions by 2005. Manufacturers vie for the governmental seal of environmental virtue, the Blue Angel, with ecologically friendly products ranging from low-polluting paints to mercury-free batteries.

Germany's pollution-prevention programs, led by its "take-back" of packaging and recycling of autos, have created a much more hospitable climate for environmental innovation and green design than yet exists in the United States. The effort, originally setting domestic environmental objectives only, also promises to pay off handsomely for German industry, which is vying for the lead in a broader spectrum of environmental technologies than either Japan or the United States. Germany is now competing with the United States in the production of devices for producing electric power from sunlight, and is likely to move ahead in the near future. Siemens is manufacturing the world's least polluting and most efficient turbines. European appliances have always been smaller and more energy-efficient than American, and new "smarter" German equipment is now increasing that gap through the use of microchips to monitor inputs and function; German manufacturers of emissions-control technology are now challenging Japanese supremacy.

Japan's drive for environmental technology, on the other hand, is a distinctly top-down affair. As a society the Japanese currently are not interested enough in other—nonhuman—species to mobilize behind popular movements to save beautiful landscapes or exotic animals and plants. Where human impacts have threatened nature, the prevailing Japanese reaction is that the harm can be forestalled or repaired by technology.

During the late 1980s, Japanese application of what environmental journalist Douglas McGill has called the "techno-fix" approach to the global environment generated some colossally wrongheaded ideas. These included proposed "superprojects," namely, the installation of what he describes as "an atmospheric control switch across the Bering Strait"; a ring of solar panels in countries surrounding the Equator, wired to cooler climes; and mammoth dams high in the Himalayas to prevent flooding in Bangladesh. At home, the same impulse has dammed and channeled virtually all rivers, churned up the landscape for development, filled in half the country's seacoast and a third of its wetlands, and "solved" garbage problems by building offshore islands of trash and incinerating 70 percent of what is not recycled.

The technological optimism inherent in the superprojects has been somewhat subdued by the recession of the early 1990s. Given the ability of Japanese society to mobilize very swiftly in pursuit of communal advantage, however, the explicit commitment of the Japanese Ministry of Trade and Industry to environmental technologies as an area of competitive growth is likely to yield significant results. They have already gained considerable competitive advantage. Originally spurred by the choking air pollution of the 1960s and the oil crises of the 1970s, the Japanese now lead both the United States and Germany in pollution technology for basic industry and in technology for energy efficiency in manufacturing and autos. Sanyo, Sharp, and Matsushita Electric Industrial now dominate the world market for solar batteries, and Fuji Electric leads in fuel cells. In 1991, Matsushita Battery Industrial commercialized the first mercury-free alkaline batteries, which it licensed to Rayovac Corporation of the U.S. Japan also produces many of the world's most advanced incinerators.

Most important, much more than businesses elsewhere, Japanese enterprises are also focusing on integrated systems that minimize waste, for example, in Nippon Steel's conversion of coal ash to zeolite for use in water treatment; and a joint project of the Japan Atomic Energy Research Institute, Chubu Electric Power Co., and Ebara Corporation to convert sulfur and nitrogen oxides into ammonium sulfate and ammonium nitrate, for use in fertilizer. The substantial ongoing support of MITI for environmental innovation and industry's habituation to energy efficiency have created a climate for green design in Japan at least as favorable as that in Germany. The ministry is coordinating research on leading-edge environmental innovations through the Research Institute of Innovative Technology for the Earth (RITE), which is drawing private-sector expertise and money into exploration of solar and geothermal technologies, fuel cells, superconductors, and electric automobiles.

In no industrialized country, however, do the economic incentives for pollution prevention and waste reduction come close to matching the rewards for after-the-fact cleanup and disposal. This is particularly true in the United States. Not only are U.S.

businesses habituated to policies fostering cheap energy and raw materials; in recent years U.S. regulations have enormously increased the profitability of waste management.

U.S. garbage-industry stock prices and profits have declined somewhat since their heady growth at the end of the 1980s, but the waste industry still averages high pretax profit margins, often over 30 percent, and net margins often exceeding 10 percent. With more than $11 billion in federal funds going for radioactive and chemical waste cleanups annually, much of it spent by the Pentagon and the Department of Energy, hazardous waste has proved a bonanza for the waste industry as well as for lawyers and consultants. During the 1980s, $20 billion was spent on emergency measures at four hundred Superfund sites and full cleanup of sixty. According to a recent report on Superfund by Resources for the Future, the hazardous waste industry will net at least $80 billion on cleanups by the year 2000, if two thousand priority Superfund sites among the estimated three hundred thousand to four hundred thousand toxic-waste sites nationwide are actually cleaned up. Pentagon and Department of Energy bills for military and nuclear cleanups are estimated at $100 billion to $200 billion, and $300 billion, respectively.

In the longer run the outsize investment we make in waste disposal may have perverse environmental consequences. While raising the costs of waste disposal increases incentives for efficient use of materials, it also increases incentives for illegal dumping. Businesses disposing of their own waste have ample opportunities to break the law. Although the enormous flows of industrial hazardous waste are monitored, business compliance with regulations is still to a great extent self-policed. Enterprises concerned about their reputations have strong incentives for compliance to avoid civil liability or criminal penalties, including jail, and to maintain their institutional images. Yet, as daily news reports of toxic dumping indicate, the temptation to cut the costs of hazardous waste disposal and its attendant paperwork also exerts strong pressure, especially on smaller businesses. Most illicit dumpers hide their tracks better than the creators of Mount Trashmore—a block-long, illegal construction and demolition dump that towered over a Bridgeport, Connecticut, neighbor-hood from 1986 to 1992—but unless they are deterred, they will continue to seek the handsome rewards of noncompliance.

The trash-hauling industry also has a long, well-known tradition of extralegal activities, including racketeering, strong-arming, smuggling, underworld connections—and of course illegal dumping. Even as giant corporations like Waste Management reap profits from stiffened regulations, the garbage industry has not put this unsavory past entirely behind it. In a not untypical example, a New Jersey garbage-hauling firm that had been charged with bribery, racketeering, and illegal dumping during the mid-1980s in 1990 opened a company called Hub Recycling Inc., whose unlicensed dump served garbage haulers controlled by two of the five New York organized-crime families. Law enforcement agencies continue to uncover garbage industry efforts to muscle out legitimate competition. Even industry leaders like Waste Management and Browning-Ferris are undergoing criminal investigations for alleged price fixing, deliberate violation of hazardous-waste regulations, and fraudulent attempts to cover up infractions.

Thus, maintaining government oversight and potential sanctions are essential to private-sector environmental performance. While businesses complain that all the regulations enacted in the past few years are ruining them, our economy continues to place increasing burdens on our environment. What should we be doing differently?

11

~~~~~~~~~~

Government Redux

I have never been more struck by the good sense and the practical judgment of the Americans than in the manner in which they elude the numberless difficulties resulting from their Federal Constitution.

—ALEXIS DE TOCQUEVILLE, *Democracy in America*

The great question of the seventies is, shall we make our peace with nature and begin to make reparations for the damage we have done to our air, our land and our water?

—RICHARD M. NIXON, State of the Union Message,
January 22, 1970

AFTER MORE THAN two decades of pollution-control efforts, it is clear that the answer to our waste crisis doesn't lie in tightening up regulations. Recently, Americans' traditional suspicion about government's meddling in their economic lives has mixed resistance with a goodly dose of disdain. Either government "can't get anything done," or when it does make something happen, the desired effect is overtaken by unintended consequences. What conservatives and liberals alike disparagingly call gridlock may represent the wish fulfillment of those who don't want government to do anything, and may even result in some measure from their efforts. In any case, in the past two decades our governors and legislators have seemed all too ready to prove their critics' case for them.

On the environment, major governmental failures can be attributed to political pulling and hauling as well as to grave institutional weakness in following through. However, they also stem in great measure from the magnitude and obduracy of the problems. Innovative initiatives have been pursued by a succession of officials for the most part apparently dedicated to their mission. Yet the gap between problems and policy has loomed progressively wider.

In an embarrassing fashion, the U.S. government itself has lagged way behind even the private sector in compliance with environmental law. Neither the Pentagon, the Department of Energy, nor the Department of the Interior was legally required to do so until 1992.* Much more serious in the longer term, however, is the great disproportion between the cumulative threats to our environment and our mechanisms for dealing with them. As environmental regulations have progressively tightened, our ability to implement them has been increasingly strained.

Municipal, state, and federal governments share responsibility and authority for handling waste. State and local governments do most of the work, regulating and supervising the collection, landfilling, incineration, and recycling of garbage; disposing of sewage; policing hazardous-waste generation and disposal; and monitoring air and water quality. They have also been responsible for much of the policy innovation on waste prevention and recycling.

Nonetheless, state programs must meet federal standards, and states rely on Congress and the EPA for guidelines and support. Over the past several decades, the federal government, in addition to establishing Superfund for hazardous-waste cleanup, has set minimum standards to regulate air, water, and soil pollution stemming from waste and its management and has provided grants to the states for research, education, and equipment. Government effectiveness at all levels depends not only on ideas and talent but on money. On both the federal and the state level, making waste and pollution policies requires adequate staffing, information, and equipment, none of which will materialize without supporting revenues. In both Washington and the states, new programs have been hobbled by serious shortfalls in all the requisite resources. In this snapshot view of implementation over the last several decades, we will focus chiefly on the federal level.

---

*In August 1992, the Supreme Court reaffirmed the immunity of federal agencies when they break environmental laws, ruling that states may not fine them, but congressional legislation passed in September 1992 ended the exemption. Various legislative compromises, however, appear likely to weaken or delay the impact of the law for some time to come.

Demands on the EPA have greatly exceeded its ability to deliver since the formation of the agency in 1970. Between 1970 and 1990, the agency's budget quadrupled and its personnel tripled, from $1.3 billion and 5,500 to $5.14 billion and 17,170. Nonetheless, the passage of a vast body of pathbreaking environmental legislation during that period increased the departmental workload exponentially. (The $5.9 billion allocated for the EPA in 1992 compared with $15.7 for the Energy Department and $61.8 for Agriculture.)

Especially in the early days of the Reagan administration, cutbacks in resources quickly eroded the competence of the EPA, and the federal government systematically and ruthlessly undermined its own ability to follow through on environmental initiatives by slashing appropriations for data gathering and oversight. Eliminating experienced personnel and substituting hired contractors who worked with little supervision, the administration proved its contention about government inefficiency: The shoddy performance of government bureaucrats and hired contractors undoubtedly did waste a substantial sum of taxpayers' money.

Environmental control enjoyed much more support in Congress. Yet in the prevalent mode of budgetary buck passing, the legislators skimped on funds for programs they had mandated the EPA to carry out and ignored fiscal obstacles when they authorized transfer of federal responsibility to inadequately funded state agencies. The gap between what Congress proposed to do and what the EPA accomplished was often enormous.

In many instances, the task assigned was patently beyond the agency's capabilities. For example, actually tracking a significant percentage of the chemicals most widely used by industry as directed in the Toxic Substances Control Act of 1976 would have cost countless billions and required a degree of EPA expertise and oversight that was certainly not available during an era of agency personnel cuts. In other instances, EPA officials appear to have construed very narrowly legislation intended much more comprehensively, as, for example, in ordering cleanups of just 529 of some 18,770 severely polluted waterways in the four years succeeding the passage of the Clean Water Act of 1987.

In part the problem arose from the adversarial relationship that developed between Congress and the EPA in the early days of the Reagan administration. After 1980, EPA implementation of previous administrations' policies slowed or halted entirely. For instance, following the 1978 ban on "nonessential" uses of ozone-depleting CFCs (including 90 percent of aerosol sprays), the agency shelved the problem of refrigerants, air conditioners, and other CFC sources for another ten years. As government pressure disappeared, industry stopped research on substitutes.

Growing congressional mistrust of the environmental agency's stonewalling led to a rigid form of micromanagement that added substantially to the agency's administrative burdens. The 1984 reauthorization of the Resource Conservation and Recovery Act (RCRA) to direct the disposition of the nation's hazardous waste illustrates the congressional penchant for dotting *i*'s, crossing *t*'s, and footnoting the fine print. The act greatly expanded the waste types and generators requiring oversight, which rose ninefold between 1980 and 1990 to 211,000, and also increased to 4,700 the number of facilities requiring regulation.* In establishing more than seventy statutory requirements, it defined technical standards in great detail, including, for example, the number and types of landfill liners. It also set tight deadlines for EPA action. Although the legislation assumed that the states would ultimately implement these regulations, Congress gave the agency an open-ended mandate to do so until the states could take over.

When the relationship between Congress and the EPA becomes a tug-of-war, the environment is probably the biggest loser. Answering their critics' charge that they cannot focus on the larger picture, agency officials point to the problems caused by legislation like the RCRA bill, citing the distractions inherent in the manifold detailed congressional directives and deadlines they must meet. Not the least of their dilemmas arises from the confusion inherent in the layering of old and new regulations. The often conflicting pressures imposed by legislation, they argue, have prevented the agency from setting priorities on the basis of environmental risk and of using available resources to address the worst sites first. Congressional activists, on the other hand, argue that without the detail and the deadlines the agency would have felt less urgency to fulfill legislated mandates. Probably both are right.

---

*The number will probably exceed 6,000 in 1994.

∽

The EPA's problems with Congress clearly intensified after the funding cut made at the outset of the Reagan administration had taken a clear toll. Although the agency's dedication to its mission of protecting the environment revived in ensuing years, its responsibilities continued greatly to outpace its capabilities. Numerous reports on environmental programs repeated the refrain: Implementation was slipshod—as a result, in great part, of insufficient resources. In 1989, for example, a General Accounting Office report on the EPA's hazardous-waste cleanup program cited gross underestimates and underbudgeting of staff time for oversight as the main factor in explaining extraordinary delays in corrective action. In March 1990 another GAO report found that the EPA's interim sludge management program, begun in 1987, had failed to meet its objectives, in large measure because of inadequate resources. In the same vein a GAO report in May 1992 pointed to the effect of budget constraints in impeding coordination of federal activities to eliminate what may be our most serious toxic risk to human health, lead poisoning.

EPA funding shortfalls delayed rule-making interminably. For example, after Congress mandated guidelines in 1976 for government procurement of recovered materials, including federal purchases of paper, construction materials, lubricating oils, retread tires, and building insulation, by 1988 the EPA had succeeded in promulgating only one guideline on construction materials. Federal regulations passed in 1984 for new municipal solid-waste landfills, which Congress required the EPA to have in place by 1988, were instead merely proposed by the agency in that year. In consequence, lawsuits brought by environmental groups sometimes seemed the only way to spring mandated regulations. The executive branch also threw an occasional spanner into the works. After the White House in 1990 prevented the EPA from issuing any regulations on protecting groundwater until it produced more evidence to show their future benefits, the regulations were finally promulgated in 1991 under pressure from a lawsuit brought by the Sierra Club, the Natural Resources Defense Council, and Friends of the Earth.

The EPA's problems in establishing institutions to protect human health from contaminated sludge are emblematic of those afflicting other underfunded agency programs. Aiming to help states set up their own administrative mechanisms for monitoring sludge, the program required the ten EPA regional offices to do the job where states failed to participate. In most cases states did not choose to take on the new burden, but neither were the regional offices able to fill the gap. Many states pointed to a lack of staff and of the computer capabilities needed to generate necessary data and oversee compliance. Obviously the one full-time EPA employee in each region responsible for managing the program could hardly pick up the slack by taking on the permitting, monitoring, reporting, education, and overall administrative detail for all the municipalities in the multistate area.

Resource shortfalls have intensified since 1980. As the GAO commented in its report on the sludge regulations, this sort of resource limitation had become a "generic problem" for environmental programs, stemming from "an increase in regulatory responsibilities arising from new legislative requirements, often to be carried out without increased federal funding." In the same period, federal support for state efforts was dwindling. By 1990, for example, most federal grants given to states under the Clean Water Act of 1972 had ended, and other sources of federal funds to state water programs had remained flat since 1981. The cumulative result amounted to a 30 percent to 40 percent cut in state purchasing power of environmental services.

As with the sludge program, dwindling revenues have generally impaired data collection and eroded staff competence. Problems with environmental data can be seen as part of a broader erosion of governmental statistical capability stemming from budget cutbacks and deregulation.* At the EPA, an advisory committee of prominent scientists concluded in the spring of 1992 that poor direction and a lack of first-rate scientists and equipment were undermining agency research and hampering efforts to reduce the scientific uncertainties that impede the creation of skillfully conceived, cost-effective regulations. The EPA research budget still had not recovered from deep cuts made under the

---

*Staffs at the Bureau of the Census, the Bureau of Economic Affairs, the Bureau of Labor Statistics, and other agencies have all been cut. Staff at the Bureau of Economic Affairs fell from 453 in 1979 to 371 in 1989.

Reagan administration. Even with recent modest increases it was still 3 percent smaller in absolute figures in 1992 than in 1981.

The results are most prominently illustrated in stalled and severely delayed assessments of risk. Although toxic waste has enjoyed a much higher priority than municipal garbage, by the spring of 1993 the EPA had still hardly begun the work mandated by the Toxic Substances Control Act (TSCA) of identifying potentially harmful chemicals. In the intervening years, the federal agency had not established criteria to define risks to health and the environment; nor had it set up an information system to keep track of test results for individual chemicals.

Failure to organize information has similarly afflicted efforts to sort out the enormous and complex Superfund cleanup tasks. Programs have foundered in the absence of a comprehensive inventory of toxic waste sites, or a program for finding new sites. Furthermore, the government cannot fully benefit from experience with ongoing programs because it lacks a centralized electronic database capable of incorporating and analyzing accumulated evidence about cleanup remedies.

Dwindling resources and declining esteem for government over the past two decades have taken their toll on staff quality. As departmental budgets have been cut, the uniquely low regard in which America holds its civil service—and the widespread characterization of public servants as idle bureaucrats—has not added to the attractions of government service for qualified personnel. Underpaid and frustrated by the gap between legislated goals and institutional capabilities at the EPA, even committed environmentalists have often burned out relatively quickly, in many instances moving into better-paying jobs in the private sector. Consequently, after twenty years of EPA work in the field, "staff inexperience" is often cited prominently among the reasons for failure in implementation.

Another obstacle to institutional competence is the preference mandated in congressional appropriations for private contractors over permanent agency staff. The argument about whether private contractors or amply funded government bureaucracies deliver common services better or cheaper has recurred perennially in America at least since the days of the Progressive reform movement early in this century. We will not revisit that debate here. Recent U.S. experience clearly demonstrates, however, that the use of private contractors is not the antidote to "ineffective government bureaucracies."

During the Reagan-Bush years, the idea that private companies could do the government's work better and for less money fostered a move to privatize not only government services but also the implementation of policy. By 1992 the amount of government spending channeled through private companies had doubled to $210 billion, about one sixth of all government spending. Although private contractors were intended to fulfill "specialized needs of a short-term or intermittent nature," they were instead often hired because agencies lacked sufficient resources to execute ongoing activities.

At the end of 1992 a study by the White House Office of Management and Budget leveled a sweeping indictment against the resulting system, wherein government agencies spent all their time hiring contractors and very little overseeing their work. Citing many billions of dollars wasted as a result, the report noted that the ranks of government auditors and other supervisory personnel had been cut at the same time that contracting had burgeoned. At the Pentagon, for example, the Defense Audit Contract Agency had 25 percent fewer employees in 1992 than in 1989. Other reports presented evidence that private contractors' purported savings actually resulted from cutting corners on quality. Frequently, however, as in the 1991 transfer to private management of the Plum Island Animal Disease Center in Orient Point, Long Island, privatization failed to lower costs as promised, even after contractors made cuts in essential services—in the case of Plum Island for firefighting staff and safety systems. A 1991 GAO report on the Energy Department estimated that department employees could have done contracted work 25 percent more cheaply.

The EPA's reliance on private contractors has been extreme. Contractors have performed most of the work at the agency (as well as at the Department of Energy and NASA). Immersed in paperwork for congressionally required reporting, overstretched EPA staff have all too seldom made it out into the field to check on contractors' methods or on the progress of their work. Per-

haps the most egregious failure of accountability, however, occurred inside the agency. The EPA appears to have been cheated
out of about $13 million in 1991 alone by a contractor hired in
the early 1970s to develop and run internal computerized information and data retrieval systems.* In 1992, the EPA even paid
a private contractor to write the official response to a congressional report criticizing the agency for relying on private contractors.

More common—very common indeed—are reports of slipshod and sometimes fraudulent performance by private firms under contract to clean up hazardous waste. Most of the billions
spent annually for Superfund cleanups go to contractors, and in
1987 the EPA inspector general himself estimated that about 30
percent of contractor costs were questionable. The following year
the GAO focused on waste endemic in the EPA-contractor relationship, asserting that the EPA was "not consistently and fully
challenging questionable contractor costs" or indicating concern
about performance delivered. The refrain sounded all too familiar in 1992, when the GAO was still criticizing EPA contracts for
providing "little incentive for contractors to control their costs,"
and noting that most EPA disbursements to contractors had not
been audited. Stating that site study and cleanup by contractors
probably has cost the government 100 to 500 percent more than
a private client would pay for the same work, a congressional
Office of Technology Assessment study in 1989 pointed to reasons why, including "redundant contractor work, poorly defined
work by the government, greater use of less experienced people,
poorly supervised work that leads to late recognition of problems,
greater concerns about being criticized which lead to unneces-

---

*For many years, the California-based Computer Sciences Corporation had
hired its own personnel to maintain sensitive information about cases against
polluters, to promulgate new policies regarding pollutants, even to review and
pay its own bills. The operation attracted EPA notice when a company employee
arrested for numerous counts of fraud against various federal programs was
found to have a prior criminal record. Early in 1992, the inspector general of
the EPA reported that the corporation, which had been operating virtually
without supervision as a kind of "private government" within the EPA, had
been double-billing for its work, charging full-time wages for employees who
worked part time, and charging the government for work never approved by
the agency.

sary, *defensive* work, and changing agency policies and personnel."

While trying to avoid an internal government bureaucracy, the system has fostered a private sector analogue "encircling" the Superfund program—working for its own permanence and expansion but not subject to public scrutiny. In a striking demonstration of just how so much slips though the cracks, *The New York Times* reported EPA officials as saying "they were so far behind in examining records of their Superfund contractors that they could not tell whether the government had paid" a company accused of bilking the agency of millions in unallowable expenses, while failing to complete cleanups.*

In addition to institutional weakness at the federal level, wide divergences among environmental regulations in various parts of the country cause different problems—though they can sometimes foster solutions. As we have seen, local action has driven environmental progress in the 1980s. Citizens' concern about nuclear waste as well as plant safety brought nuclear energy development to a halt. Local resistance to new landfills and incinerators and pressure for recycling clearly have pushed municipalities and industry into materials recovery and laws supporting waste reduction.

California's pioneering environmental legislation is a prime example of progress growing out of pluralism. The peculiar severity of Southern California's mountain-trapped air pollution as

---

*A sideshow here is the recent uproar over interstate garbage shipments. In 1992, the restriction of interstate garbage shipments emerged at the political frontline of the garbage crisis. An increasing volume of America's trash, 7 or 8 percent in 1989, now around 10 percent, rides the road in pursuit of cheaper dumping fees. Unpopular almost everywhere—except with sanitation officials facing high landfill costs—trucking garbage consumes energy, litters highways, and inundates communities that host landfills. Opposition to it ranges across the political spectrum. In 1992, the Supreme Court ruled that the interstate commerce clause prohibits states from restricting imports and exports of garbage. On the other hand, in the same year, legislation passed in the House that gave states new authority to restrict garbage imports was probably the one piece of environmental legislation that enjoyed overwhelming congressional support. Meanwhile, vast quantities of garbage continue to ride the road.

well as the state's scanty water supplies has spurred a uniquely urgent response, and other states with less extreme natural problems have followed suit. So, eventually, has the federal government. California's environmental law sets the pace on air pollution regulation, not only in the United States but for the rest of the Western world as well. California legislation mandating that electric cars constitute 20 percent of the sales of each automotive manufacturer by the year 2000, for example, has quickened competition among vehicle manufacturers in the United States, Japan, and Europe.

Where both the economic and the environmental costs and benefits are more ambiguous, however, and where the political winds are less robust, a cacophony of individual efforts often yields confusion and frustration rather than progress. The legislation that the German Bundestag is using to force conservation in packaging and manufacturing will not work the same way if passed by the City Council of Minneapolis, or Long Island's Suffolk County, or even the state of Massachusetts. The packaging industry is unlikely to shift from selling more materials to providing more environmentally efficient products without enormous external pressure, and the change will entail substantial transitional costs. Waste minimization via the German packaging law, for example, costs twice as much in the initial stages as the conventional waste disposal methods it is replacing. No municipality or state by itself could administer such a shock to the system.

In 1992 the state of Minnesota instituted a campaign to educate consumers about the outsize costs of highly packaged products and their outsize contribution to the waste stream, by means of dramatic comparisons, posted in stores, schools, and libraries, of, say, a five-pound bag of rice with a boil-in-bag box, or a tube of toothpaste with a pump dispenser.* As creative as the promotion and the comparisons may have been in reaching shop-

---

*A five-pound bag of rice costs 47 percent less per unit than a boil-in-bag box, which leaves 98 percent more waste; a 6.4-ounce tube of toothpaste costs 36 percent less than a 4.6-ounce pump, which leaves 69 percent more waste; a one-gallon plastic jug of spring water costs 87 percent less than six twelve-ounce bottles, which leave 83 percent more waste; a 5.9-ounce box of instant pudding costs 64 percent less than a ready-made snack six-pack, which leaves 89 percent more waste.

pers, their effect on how products were actually wrapped and displayed was extremely modest when compared with the force of the direct pressures changing industrial behavior as a result of the German packaging law. It would probably be quite a long while before cash receipts from Minnesota registered loud and clear in corporate headquarters elsewhere.

The uncertainties for businesses operating in a number of different jurisdictions also impede effective response to environmental problems. For large businesses, the costs of complying with diverse and shifting regulations usually do not threaten survival, but they do drain management time and resources. Most serious, both for the environment and for national competitiveness, they contribute to the prevailing climate of defensiveness throughout the private sector about dealing with waste and pollution. For small businesses with modest marketing, distribution, and legal budgets, the cost of compliance with a patchwork of environmental regulations bears down particularly hard. Even for enterprises operating entirely within one state, permitting requirements by separate federal, state, municipal, and local governments can multiply bureaucratic burdens. Where business crosses state lines, difficulties increase.

According to an article in *The Wall Street Journal,* for example, an Auburn, Washington, stove manufacturer that had waited to design its combustion system until the EPA set an emissions standard for particulate matter of 7.5 grams per hour had to spend $250,000 to revamp all its stoves after Washington's state legislature lowered the permissible level to 4.5 grams. A Pennsylvania manufacturer of water purifiers ran into trouble in three states: Iowa, where that state's regulators threatened its exclusion from mail-order catalogs nationwide after the company decided not to submit to Iowa's stiff and costly registration procedures; California, where the company spent years trying to prove water purification claims for the product; and New York, where a regulation that "purifiers" must remove all viruses impelled the company to change its product's name to a "microfiltration system" and redo all its literature for the New York boat show.

While most managers will argue vociferously against tightening standards, many will admit that their worst problem is uncertainty. They need harmonization of policy, necessarily at the federal level, to provide a predictable context for planning. Of course businesses hope that central coordination may produce the least common denominator, allowing them to sidestep the Californias, Minnesotas, Iowas, and New Yorks. Yet at a more fundamental level, they crave a stable environment in which to adapt to the standards society defines as optimal. In 1991, while lobbying for the White House to ease restrictions on development of wetlands, many developers acknowledged that changing the rules for the third time since 1987 would probably cause further delays and alterations in the design of projects, as state and federal governments battled over how the regulations would be applied.

In addition to the jurisdictional complexities, the American regulatory approach itself compounds the confusion. When we embarked on our pioneering effort to control pollution at the end of the 1960s, we were undertaking something akin to a revolution. In a spate of rule-making that shifted the locus of regulation from the states to the federal level, we attacked the growing contamination of air, rivers, and oceans. The controls we erected then and thereafter have been essential in slowing the pace of deterioration. Yet they satisfy neither environmentalists nor business nor the Congress that enacted them. In fact, they don't work very well.

There is no doubt that we need governmentally enforced standards. The weakness of purely voluntary guidelines has been thoroughly demonstrated. In Arkansas, when President Clinton was governor, prescribed behavior to prevent contamination of rural water systems appears to have been largely ignored by farmers on many thousands of acres. In the upshot, eroding topsoil and wastes from over a billion livestock left watersheds degraded and miles of streams threatened by silt and slime.

For industry, increasingly strict standards for hazardous waste have started to change the way businesses view the "by-products" of their operations. Together with Superfund liability provisions, these regulations have started to create a climate in which the effect of wastes on the bottom line is beginning to be taken more seriously. By increasing the costs of waste disposal and cleanup the new rules have multiplied pressures to curb toxic emissions.

The Toxic Release Inventory pollution reporting system has further raised industrial consciousness, both from fear of public exposure and from managers' shock when they themselves recognize for the first time the total spectrum of pollution associated with their activities.

Regulations may also drive advances in industrial design, bringing innovations in pollution prevention or recycling, as in California. Germany's regulatory innovation in mandating manufacturers' "take-back" of their products for disposal is changing the structure of the market, greatly increasing incentives for "green design." It is doubtful that German auto manufacturers would have launched current initiatives in recycling and reusing automobiles and electronic goods if the rather radical German laws had not been passed.

Over the last several decades, environmental standards have evolved in the direction of somewhat greater efficiency. Early efforts to mitigate pollution focused on "end-of-the-pipe" pollution controls, and regulations prescribed "the best practicable" or "best available" control technologies for each problem. Then, as concerns broadened to include the whole range of toxic substances flowing from myriad sources into air, water, and land, attention shifted from technologies to emission levels. While setting specific levels according to perceived health risks, later regulation allowed operators to choose the technologies more freely to achieve these standards.

During this period, it has become evident that rules specifying outcomes to be achieved are usually more efficient than those prescribing the technologies to achieve them. Generally, detailed directives decrease the flexibility of industry responses and thereby dampen innovation. Once businesses invest in a prescribed set of devices to cut emissions at the end of the pipe they are less likely to look for changes in their processes or products that will reduce toxics at the source. A steel mill that was forced to install an elaborate filter system to contain gas and metal emissions would probably be disinclined to invest in research on ways to eliminate pollutants in the combustion cycle. It would also be more likely to strongly resist raising standards, on the supposition that the new rules would consign its current emissions control technologies to obsolescence.

As standards tighten, however, no sort of regulation looks good to industry. When pollution control reaches a certain point, the costs of eliminating additional emissions generally rise sharply. Industry has responded to increasingly tough laws by investing in resisting regulation. As EPA studies indicate, in the absence of extensive oversight, the self-reporting mechanisms that monitor compliance with many U.S. environmental regulations fall far short of providing an accurate picture. In the face of tightening regulations on waste disposal, some businesses choose to dump illegally. Many others confront swiftly changing, sometimes conflicting regulatory directives by lobbying, keeping their heads down, or litigating.

Pursuing higher levels of environmental control through command-and-control regulatory systems is likely to exacerbate systemic problems. Placing greater demands on administrators will weaken enforcement; inflexibility will impede the evolution of technological innovation; and compliance costs for industry will rise significantly. The GAO estimates that costs of controlling pollution will rise from $115 billion in 1990 to around $160 billion in the year 2000. At the same time, the EPA's budget requests for implementation of mandated pollution-control standards are falling relative to increasing burdens.

In addition, as we have seen, the American regulatory system creates peculiar difficulties for implementation. Compared to procedures followed in the other advanced industrialized nations, the American way of setting up, implementing, and enforcing environmental controls exacts inordinate costs of both government and business, without appreciably increasing benefits. Our pervasive wariness of government intervention, the notion of checks and balances ruling relations among the branches of government, and the legitimacy of continued contention among the various parties all play a role in complicating our version of environmental regulation.

Because American regulatory legislation encourages resort to the judiciary, Americans challenge the rules in court far more frequently than do the Europeans and Japanese. After lobbying the legislature, businesses and environmental advocates continue to pursue their ends by other means—through lawyers, depositions, injunctions, and costly suits. The price to our economy looms even higher when the entrepreneurial energy and initiative wasted in the adversarial process are added to the billions spent on litigation.

The costs of judicial second-guessing can also be high. A decision by a federal judge in January 1993 that blocked new automobile emissions standards for New York State threatened to snarl the state's pollution-control processes and curb access to federal money for enforcement of regulations. If his finding an undue burden on automakers is upheld in a new trial in 1994, the judge's decision will force the state to resort to other ways of meeting air-quality standards, possibly through what state officials called "draconian" restrictions on industry and curbs on traffic. New York will also have to embark on a new regulatory process, inviting intervention by legislators and other interested parties, and increasing odds against meeting federal standards. Failure to meet those standards would subject the state to serious sanctions including loss of billions of dollars in highway funds.

In part because business resistance to regulation is assumed by all parties, and in part because of congressional efforts to keep a tight rein on executive agencies, regulatory measures involving complex micromanagement of procedures are followed by extensive reporting requirements for business. At various stages, studies must be made, permits acquired, and forms filled out. Separate permitting requirements for federal, state, municipal, and local jurisdictions increase the bureaucratic burden, as does overlap between statutes passed successively to deal with similar problems. The structure of responsibility at the EPA and other environmental agencies, which separates oversight of air, water, and land, adds to the confusion. To illustrate the dimensions that paperwork can assume, *The Wall Street Journal* described a photo of an employee of a North Carolina furniture plant standing beside a stack of forms dealing with disposal of the plant's dirty cleaning rags: The pile of paper topped the man's six feet, two inches.

Recent efforts to create a less adversarial process, whereby interested parties could shape regulations through negotiation, have not gotten very far. In the American context, regulatory negotiation, whether established through laws (the Negotiated Rulemaking Act of 1990) or essayed in a few notable standard-setting exercises, has been less than a resounding success. In 1990, potential industry support for a less adversarial approach did surface during EPA-organized negotiations on airborne toxic leaks. In those talks, the chemical and petroleum manufacturers devised incentive schemes for penalizing noncomplying plants and rewarding those that exceeded standards, which made possible unprecedentedly tough limits on a large number of chemicals. Environmental groups have also begun recently to work directly with industry on particular problems, most notably in the Environmental Defense Fund's collaboration with McDonald's, starting in 1990. The victories for negotiation, however, represent tiny islands in the roiling sea of litigation.* That the idea of a consensual process has yet to take firm hold is suggested by the behavior of the ethanol industry: In 1992, after the completion of successful negotiations on standards for cleaner-burning gasoline and after it had signed off on the regulations, the industry immediately began seeking a special loophole allowing more smog-forming emissions for each gallon of ethanol than for other types of gas.

In the environmental arena, Superfund epitomizes the confusions and waste inherent in our current regulatory approach. The regulation's retroactive extension of the "polluter pays" principle to parties long gone from the site has indeed erected a strong deterrent against careless dumping of hazardous waste, or any wastes that might be viewed as hazardous in the future. European governments, convinced of the effectiveness of retroactive sanctions, are currently exploring adoption of modified versions of the American law. In this country, however, Superfund regulations have encouraged a surge of legal actions among all the parties potentially sharing responsibility.

Superfund liability provisions can reach to almost any company remotely involved in the site—those who sent waste there, those who transported it, those who disposed of it on the site— regardless of the presence or absence of alleged negligence. Be-

---

*Those with deep pockets necessarily enjoy an advantage. In the past few years, industry has effectively employed the threat of suit (known as Strategic Lawsuits Against Public Participation or SLAPP suits) to deter individual citizens presumably with few means to defend themselves in court from opposing development projects.

cause the program has relied on the courts to force recalcitrant polluters to clean up after themselves or reimburse the EPA for doing so, perhaps half the agency's money has been spent searching for responsible parties, negotiating settlements with them, and paying for litigation and judicial enforcement when negotiations fail. In numerous cases the responsible parties are dead, bankrupt, or nowhere to be found. Thus the public costs have been high, even if some portion of them have eventually been recovered.

Superfund offers the most extreme example of the exorbitant money costs of environmental litigation. A study by the Rand Corporation found that from 1986 to 1989, 88 percent of insurance company payments on Superfund claims went for "transaction costs," roughly half of these to dispute claims and the other half for policyholders' legal fees. With 30 percent to 60 percent of the billions spent so far for Superfund going for adversarial contests among potentially responsible parties (PRPs) and between PRPs and the government, experts estimate that legal costs for Superfund alone could top $200 billion before the job is done.

Study and permitting requirements involved in identifying and planning Superfund remediation have also slowed down cleanups. Remediating a typical toxic waste site now takes twelve years from start to finish, at a cost of more than $30 million. While exhaustive review and reporting on remedial plans has gone forward, actual work on eliminating contamination lags behind. By the end of 1993, as we have seen, after an expenditure of over $25 billion, only 57 sites had been deleted from the list.

Nor has the regulatory process aided in sorting out conflicting priorities. Rather, the rigidity of the rules had increased the contradictions between conflicting goals. For example, RCRA definitions of waste, which require extensive permitting and documentation of industrial by-products, have sometimes impeded pollution prevention, recycling, and reuse. Regulations requiring wastes containing even small amounts of toxic materials to be defined as hazardous, designed to prevent the mixing of hazardous wastes with other substances as a sort of camouflage, may in some instances act as a disincentive to responsible disposal or recycling. For example, steel makers contend that the extraction and reuse of lead, zinc, and other heavy metals from furnace ash becomes uneconomic when the residue must still be disposed of as hazardous waste. Rules requiring used oil to be handled as hazardous waste sometimes impede recycling and probably lead to illegal dumping.*

Responding to the inflexibility of the Superfund regulations and the costly litigation arising in their wake, the Clinton administration, in early 1994, proposed the new approach outlined in Chapter 5. While easing cleanup requirements somewhat for areas not intended for residential use, the proposed changes would also encourage responsible parties to submit to arbitration, by forcing those who refused to settle to pay a higher share. If enacted, the measure appears likely to rationalize remediation costs. Whether it will effectively curb polluting waste, however, is less clear. In general, loosening pressures for compliance correspondingly increases the need for oversight.

Beyond the regulatory maze, American environmental efforts have also suffered from faltering government initiatives in support of the necessary intellectual and physical infrastructure that we all depend on but no one owns. In the early days of the Clinton Administration, "public spending" projects seemed to be regarded on the one hand as a convenient way to infuse money into the economy, and on the other as a species of pork-barrel boondoggle. Both perspectives slighted the urgent need to repair and renew a national infrastructure fallen prey to the quest for small government in a time of generally declining rev-

---

*Alcoa environmental manager R. Lee Byers cites numerous instances in which RCRA regulations have posed barriers to pollution prevention. For example, rules distinguishing between raw materials and solid waste forced cement manufacturers to terminate use of spent potlining residues from aluminum smelting as a source of fluoride for cement. A technology to thermally oxidize chemical wastes to recover both energy and chlorine in the form of hydrochloric acid was abandoned in the face of the two- to six-year permitting time and the corrective-action investigation that would have been imposed on an entire plant. Many secondary materials must be purified before they can be used again, but the need for exhaustive RCRA permitting for materials stored before processing often raises costs prohibitively. The burning of wastes generated in a petroleum refinery for energy in a catalytic cracking unit requires the entire unit to be permitted as an RCRA treatment facility.

enues. American spending on roads, bridges, mass transit, airports, school buildings, and telecommunications fell from about 4 percent of GNP in the 1950s and 1960s to about 2 percent in the 1980s, two to three times lower than that of our principal competitors, Germany and Japan.

While private sector R&D budgets are falling, government support for R&D to spur technological innovation, particularly on the environment, is also lagging. In 1992, U.S. government support for industrial R&D was minuscule—less than 1 percent of all government R&D investment (which was 50 percent less in absolute terms than Japan spent). As defense labs scale back, falling federal support for science constrains research options more generally. Further, existing U.S. government support for technology continues to accord very low priority to the environment.

As the World Resources Institute economists have pointed out, this shortfall results partly from the decentralized nature of technology policy in this country. Unlike most other industrialized nations, the United States has no national department of research and technology. Nor do we support industrial technology generally, but rather channel R&D monies into separate efforts on defense, agriculture, health, and energy as well as the space program. Within those specific missions, the environment has not fared well. In 1989, for example, while Star Wars received $6 billion, the total energy-research budget came to $500 million. We have seen the impoverished status of alternative, environmentally sustainable agriculture, at around 1 percent of public agriculture research budgets.

Total U.S. government-supported environmental R&D amounts to around $4.8 billion, or 7 percent of total government R&D expenditures. Although higher in absolute terms than that of Japan or Europe, a large proportion of the U.S. total represents investment in environmental-control and waste-disposal technologies, including emissions controls and monitoring, incineration, and landfilling. The Japanese and Germans, on the other hand, are investing significant amounts in innovative process technologies for reducing waste and conserving materials and energy. New products and processes flowing from the Japanese "eco-factory" concept, for example, aim at fulfilling future needs for environmental efficiency rather than at complying with present regulatory mandates. Given the continuation of these trajectories, the United States is likely to lose further competitive edge in the environmental area.

In moving toward a conservationist economy, the United States has to play to its own strongest suit. We could not imitate Europe or Japan even if we wished to. We can, however, use government policy to accelerate progress, and perhaps simplify process. Given our prevailing individualism, we need, even more than most of our chief partners and competitors, to use the market as an organizing tool. But to do that successfully, we also need to get the incentives right. In the words of State Department counselor and former Senator Tim Wirth, we must "put a green thumb on Adam Smith's hand."

12

≈≈≈≈≈≈≈

Policy, the
Environment, and the
Marketplace

*Society is always taken by surprise at any new example of
common sense.*

—RALPH WALDO EMERSON (1803–1892)

What do we do now, in order that we may get to the next step?
—DWIGHT D. EISENHOWER, on conservation, 1954

WHETHER WE CALL it poor housekeeping or a stupen-
dous accounting error, our failure to value our air, land,
and water appropriately has left us with an extremely costly
mess. Contending with the damage already done and forestall-
ing more of it appears to be such an expensive proposition that
many people ask whether it's worth what we would spend. They
would not choose to pour money into "the environment" if it
meant trading off significant amounts of income now used for
other things.

The choice between economic and environmental goods, however, is a false one. The idea that protecting the environment
hurts the economy seems to make sense if we look at environmental protection chiefly as development foregone, or if we
merely count up the costs of pollution control. However, this sort
of calculation leaves out more than it includes, such as safeguards
for health; conservation of resources and protection of species
for future consumption; preservation of natural beauty as a basis

for tourism; savings by industry from reductions in materials inputs and wastes; and both jobs and profits flowing from new services predicated on conservation of resources. In 1992, the U.S. environmental-services industry counted nearly seventy thousand businesses, with an income of $134 billion, employing over a million workers.

Moving from wasting our environment to sustaining it,* however, will require much more purposeful restructuring of policy than Americans ordinarily engage in. Can we bring ourselves to do it? After surveying all the activity devoted to coping with waste over the past few years—and the real change in people's attitudes—it is clear that we have come some distance but are still very far from where we should be.

In the case of toxic emissions, although things are generally getting worse, not better, in 1994 cities, states and perhaps the federal government, concerned about burdens on business, were easing regulations on disposal of hazardous wastes. Throughout the environment, many millions of tons of toxic chemicals continue to be released each year. As population and economic activity grow, both municipal and industrial waste continue to mount, as do absolute levels of most pollutants. Even as we make some progress on a few major targets, like lead, others get out of hand. Yet many city governments are moving back from stringent rules mandating expensive cleanups of contaminated sites in order to attract businesses to developed urban areas. Thus they are weakening deterrents to toxic waste generation.

Most of our efforts at present still go into what has been called "the environmental protection paradigm." We keep busy managing waste and containing pollution, around landfills, at the end of smokestacks and tailpipes, at industrial and sewage outflows into waterways, from ships' bilges and leaking tankers, around

---

*The term "sustainable development" came into common parlance in the Bruntland Commission report on environment and development of 1987. Defined by the commission as "development that meets the needs of the present without compromising the ability of future generations to meet their own needs," the phrase was also heavily used in connection with the 1992 Rio conference on the global environment. Though its currency deprives the words of much rhetorical impact, the subliminal message of the oft-repeated "sustainable" works toward a much needed change in attitudes, underlining the value of maintenance and preservation.

nuclear repositories. In addition, we seek, rather ineffectually, to clean up the worst of our past excesses.

We have made only very modest progress in moving from waste management to reduction. While environmentalists, packagers, and municipal officials debate whether or not recycling is "cost-effective," budget deficits jeopardize any spending on new programs. Recurrent outlays for waste reduction continue to be very modest in comparison with expenditures on other alternatives for dealing with waste. Despite the surging public enthusiasm for recycling, a 1989 survey of eighteen Northeastern states found that planned state spending on incineration from 1989 to 1995 was eight to ten times higher than that for recycling. Both federal and state budgets for recycling are still small in relation to overall waste management appropriations, and in 1992 many of the larger, more established programs were falling prey to recessionary budget cuts. In 1992, for example, Illinois moved $7.8 million from the state's solid waste management and recycling fund, generated from landfill surcharges, to its general fund.

Nor do sales of recycled commodities seem likely to make materials processing a reliable source of revenue any time soon. Despite the positive recent initiative by the Clinton administration, mandating that all paper procured by the executive branch contain 20 percent postconsumer recycled content by 1995, federal procurement mandates have only begun to materialize. In fact, many federal procurement policies, particularly where military specifications come into play, call for use of virgin materials, CFC-containing cleaners, and leaded paints, even where these materials are not necessary for product performance. While the patchwork of state procurement standards, landfill bans, some tax relief, and consumer codes is starting to be felt, markets for most recyclables are still small and fragile. Usually source reduction has been positioned as a "minor afterthought" to recycling.

While federal legislation on hazardous waste has induced a new caution in industry about disposal of wastes now defined as toxic, and also spurred some pollution prevention initiatives, these have been small in relation to the size of the problem. In 1989, only 1 percent of the $80 billion spent on the environment went for pollution prevention. The need to report on releases of

320 chemicals to the EPA's Toxic Release Inventory (TRI) has brought significant decreases in airborne emissions reported by larger U.S. industrial facilities, but minimal drops in land releases and none in water. Many facilities, of course, are not required to report emissions. Moreover, industry at present is required to report only on toxic materials that are not yet incorporated into products. A truly comprehensive picture would include the zinc mixed into paint, the lead in batteries, the chlorine in paper, as well as materials lost or discarded in the manufacture of these goods. Adequately measuring toxics use, however, would be a huge undertaking, meaning greatly increased paperwork.

To speed environmental progress we will have to enlist market forces. Before we can do so, however, we need to understand how our current method for calculating economic welfare distorts incentives. The first problem lies in the fact that our economy still incorporates systemic supports for exploitation of resources. Nominal fees for use of public grazing lands have decimated our rangelands; tax abatements favor new tract housing over renewal and renovation; government insurance supports development on floodplains and coastal shores; the Department of Agriculture subsidizes farming that requires heavy use of fertilizer and pesticides; minerals and timber are to be had merely for the taking; and energy prices rank among the world's lowest. All these policies encourage our economy to dispense materials liberally throughout the environment.*

---

*Financial structures also skew the way we calculate costs of dealing with garbage. Typical incinerator contracts include "put or pay" agreements guaranteeing delivery of certain volumes of waste, provisions passing financial risks on to the municipality, high dumping or "tipping" fees, favorable contracts for electricity, large state grants, and often free land on which to build the plant. Local governments frequently absorb site development expenses for incinerator plants, including the costs of extending water and sewer lines and bringing in roads. Similarly, in costing landfills, government estimates seldom include closure and liability expenditures in charges to current users.

Few recycling programs enjoy these same implicit subsidies. Given the assumption that collected materials will be sold, recycling programs are still expected to cost less than incineration or landfilling. Estimates usually obscure avoided costs for both collection and disposal, and sometimes governments fail to seize these savings, as when they allow garbage carters to resist cuts in regular collections and to treat recycling as an add-on. Major investment banks continue reluctant to issue revenue bonds to support large materials-processing facilities. Similarly, municipal managers have been slow to structure revenue bonds for financing recycling facilities with guaranteed repayment through mu-

In the competition between virgin and recycled materials, subsidized energy is essential for the extraction of minerals from the earth as well as in processing them. Energy breaks subsidize the primary production of aluminum by about 17 percent of the delivered cost of the metal. The story is similar for other materials. Paper manufacturing ranks fourth in industrial energy use in the United States. In 1988, virgin paper producers benefited from an energy subsidy ranging somewhere between $262 million and $347 million, about 2 percent of the overall cost of materials.

Even more important than outright supports for resource exploitation, however, is the underlying economic theory that encourages it. This paradigm virtually ignores depletion of resources—and indeed the material environment. World Bank economist Herman Daly points out that "a search through the indexes of three leading textbooks in macroeconomics reveals no entries under any of the following subjects: 'environment,' 'natural resources,' 'pollution,' 'depletion.' One of the three does have an entry under 'resources,' but the discussion refers only to labor and capital, which along with efficiency, are listed as the causes of growth in GNP." (The texts cited include R. Dornbusch and S. Fischer, *Macroeconomics* [1987]; R. E. Hall and J. B. Taylor, *Macroeconomics* [1988]; and R. E. Barro, *Macroeconomics* [1987].) Economists generally view the environment not as part of macroeconomics, which treats larger economic aggregates, such as employment or inflation, but in terms of "microeconomics," which analyzes smaller elements, including the behavior of individual consumers and firms. Thus, a somewhat fairer test from the economist's perspective would be a textbook providing an overview of macro- and microeconomics, such as Paul Samuelson and William Nordhaus's *Economics*. The index of this best-selling basic text lists four pages under "environmental damage" and the same four pages plus three more under "pollution." Except for several other scattered paragraphs treating environmental issues, those few pages appear to sum up the subject for this 950-page work.

---

nicipal tipping fees to the plant operator. Thus although various mechanisms for funding recycling have been institutionalized, it is still often viewed as a sort of luxury, compared to landfilling and incineration, undertaken to make people feel good "as long as it doesn't cost too much."

Appropriating metaphors from Newtonian physics, the late-nineteenth-century economists whose thinking still shapes mainstream economics today tidied up theories about human production by excluding the natural world. As Daly argues most powerfully, in neoclassical economics the physical basis of production in energy, soil, water, wood, and metal simply vanished. Along the way, technology was assumed to be infinitely substitutable for resources.

Time also disappeared, as economists firmly subordinated the future to the present. In the abstract flow of production and consumption between firms and households, value was deemed to reside in the immediate preferences of the individual consumer. Since most people when given a choice opt for a bird in the hand over two in the bush, the present was presumed to rate above the future in most "rational" calculations. This way of computing future values ensures that on paper they will invariably be less than present worth. People can always gain interest on the money earned by selling a property, so its future value is "discounted" by subtracting a given percentage of present value compounded by the number of intervening years. Although economists apparently believe the free play of individual initiative through the market will make things better and better, "discounting for the future" ensures that the welfare of succeeding generations is always subordinated to present preferences. Exploitation now is always preferred to conserving for later use.

The problem with the economists' extrapolation from people's marketplace behavior is, first of all, that the choices measured reflect a narrow selection from the range of values that people should and do consider in making decisions for themselves and their progeny, for the present and the future. Secondly, many economists have in effect turned observations about behavior—what *is*—into norms, or what *must* be. Thus, even if, given growing scarcity, a stand of mahogany might be worth five times its current value in the near future, by ordinary economic calculations it ought to be cut down now.

This calculation of future worth makes it difficult to justify in economic terms preserving resources for future use. If fishermen harvested all representatives of a declining tuna population in one season, they might deposit significant sums in municipal funds when they were done, which would accrue a lot of interest.

Nonetheless, while those few fishermen were enjoying their wealth, no one would eat fresh tuna the next year or ever after. For reasons that have become more obvious recently, the failure to calculate whether the materials we use are sustainable under current conditions of production and consumption threatens the continued viability of our air, land, and water as well as minerals, supplies, and animal and plant species. Thus it puts all our economic efforts at risk.

Sustainability of so-called "renewable" resources may be as dicey as limiting use of "nonrenewables." We know that exploiting nonrenewable minerals like oil, iron, and aluminum heavily now means certain scarcity later, even with recycling, unless substitutes are found. Thus rationing may be necessary until we find alternatives. This prospect, which can be borne with relative equanimity by technologically advanced societies in the United States or Japan, may well spell impoverishment for primary producers like the Malaysians or Nigerians. On the other hand, commodities we had thought infinitely renewable, like fish and plants, or seemingly omnipresent environmental assets, like air, water, and soil, can regenerate themselves only up to a certain point. Thus, we now begin to see that overly intense exploitation of these renewables may also cause irreversible depletion.

Given the neoclassical theorists' world, it is not surprising that they left waste out of their calculations. Over a hundred years ago, amid the explosion of early industrial development, natural resources paled in importance beside machines like the steam engine and the capital to produce them. Nor could the economists of the day have foreseen how the enormous growth in scale of human productive activity would mark this earthly habitat. In part, however, this particular interpretation of economic reality grew out of their choice of metaphors. The need to accommodate theories appropriated from mechanics may have driven their decision to scant value of energy and natural resources. The theories and the accompanying mathematical equations wouldn't have worked otherwise.

In fact, they don't work. Lacking complete capital accounts and proper balance sheets, in the real world this approach distorts and sometimes reverses calculations of costs and benefits. As we struggle to keep from choking on the residues of production, economists still use a measure of "consumption" as our main gauge of welfare. At the same time, when they mention environmental assets and costs at all, they define them as "externalities"—as outside the system. Although air, water, soils, and natural resources are used in producing what we consume, unlike plant and equipment their deterioration or depletion are not counted as depreciation. Instead, government expenditures on mopping up toxics or treating lead poisoning are counted as growth. Consequently, real losses in national wealth are misrepresented as increases in income.

Like the conviction held by people in medieval times that their planet, Earth, was at the center of the solar system, our prevailing images of our economy overstate the importance of human constructs. The idea that environmental assets and costs are "externalities" trailing off at the edge of the economic arena has the tail wagging the dog. The human economy is a subset of the environment, not the reverse.

Although a few younger economists are vigorously questioning the neoclassical approach, it has become deeply embedded in our national psyche. Most people don't think very much about the workings of the market economy, but they are bombarded with statistics reflecting the neoclassical equation of welfare with production and consumption. The national income accounts have made the quarterly growth of that product the most important measure of domestic performance and well-being. Based on monetary values—prices of goods and costs of producing them—the figures draw power from their concreteness and simplicity. As the nineteenth-century British physicist Lord Kelvin once remarked, "When you cannot measure it, when you cannot express it in numbers, your knowledge is of a meager and unsatisfactory kind." Although the National Accounts were invented fairly recently, in 1947 or thereabouts,* by now their message has taken on the quality of Holy Writ. Forecasting the quarterly estimates has become a hundred-million-dollar industry.

Capitalist organizing principles are similarly embedded, though perhaps less intensely held, in European worldviews and in a somewhat altered Japanese variant. As embodied in national accounting systems they also shape the mind-sets of finance min-

---

*Somewhat differing estimates of the U.S. national income and its components had been made by the Commerce Department since the mid-1930s.

istry officials everywhere. The U.N. equivalent to the U.S. national income account, the System of National Accounts (SNA), is in use in various national guises throughout the world, as well as in the accounting systems of other multilateral institutions like the World Bank and IMF, the EC, and the OECD. For many Third World peoples, however, existing largely outside the orbit of Western culture and education, these ideas are nearly as alien as they would have been to our own forebears up to the nineteenth century.*

To get an accurate picture of how we're doing, we must revise the way we do our sums. If we mean to change the relationship between the human economy and the natural environment we must alter the way we calculate national income and wealth. Until our national accounts factor in natural resource flows and environmental costs and benefits they will continue to seriously mislead us about both our economic prospects and our general welfare. Efforts to adapt the U.N. System of National Accounts and those of most other industrial nations have been underway for many years. Partly for lack of impetus, and partly because the problems are so complicated, progress has been limited.

Admittedly, it is easier to target the anomalies than to correct them. It is not difficult to understand how leaving out the costs of natural resource depletion leads to inflated estimates of national well-being. For example, the costs of acid rain in terms of declining fish populations, dying forests, and blighted plant life are not subtracted from income until factories purchase scrubbers or other mandated technologies. Where government or citizens, rather than the polluter, pay for cleaning up the mess, environmental costs are counted as income. The same holds in the case of resource extraction, say, mining or fishing, where the consumption of natural capital in the form of steel or oysters is counted as generation of income and not as depletion or dete-

---

*Critiques of the national accounts also come from various other directions, arguing for inclusion of: housewives' services and other household activity and unpaid labor; leisure activities; changes in quality of life; intangible capital resulting from research and development and from education, child-rearing, and job training.

rioration. Further, once we have taken oil out of the ground we count up the income it generates but neglect to subtract the value of the stock used up—while we do, however, subtract the depreciation of rigs and drills. Thus the failure of national income accounts to allow for the depreciation of natural resources exaggerates national product.

Although these problems are obvious, it is less clear what we should do instead. Strategies for correcting our calculations fall into two broad categories: establishing monetary values and measuring physical volumes. In the latter approach, flows of materials are tracked to aggregate raw materials used and wastes emitted in both production and consumption, counting up, for example, tons of vegetable matter lost and of carbon released, acres of forests cut and soil eroded, and rising and falling water-quality indices. Known as "reconciliation," or "satellite," accounts, these tabulations are not supposed to supplant but to supplement the main accounts, furnishing a physical analogue to models of production and consumption.

It will be difficult to change the way we think about national accounting without establishing monetary accounts, yet doing so presents obdurate problems. Since much of what must be valued is currently provided by nature or subsidized by the government, it does not receive a monetary value in the marketplace. Clearly, commodities like minerals, timber, and fish, which are bought and sold, are much easier to factor into the accounts than "free goods" like air and water. For this reason, most progress has occurred in estimating the effect on national wealth—or sustainability of income generation—of exploiting those resources.

On the other hand, efforts to estimate the marketplace value of environmental quality—either what is lost with increasing pollution or what is added to income and/or well-being by environmental assets—have been halting. Some of the costs of pollution are already counted in the pollution control expenditures of industry. Another method proposed for capturing costs is counting "defensive" expenditures for pollution control by government, for example, on purifying contaminated water, treating pollution-related illness, and monitoring the toxicity of chemicals. Alternative approaches include assessing lost productivity, gauging citizens' "willingness to pay" for environmental benefits, and estimating how much we would spend restoring contaminated areas to an unsullied state. Current heated disputes over "contingent value" measurements, which ask people what they would pay to protect a wild animal or keep the oceans pure, demonstrate the difficulties. None of these assessments are easy, nor do they fit into the current overall system.

With regard to stocks of environmental assets or improvements in environmental quality, it is particularly difficult to establish a standard of comparison or to disentangle "economic" payoffs from more general "welfare" benefits. How, for example, does one count the pristine quality of an ocean bay—in terms of the value of the oysters spawning there, the health of the children bathing there, the profits of the inns lodging vacationers, the pleasure of the sailfishers skimming across the water? In the Hudson River, does the market value of the returning striped bass and other fish, or the rising price of shorefront real estate, express more than a fraction of the gains from reduced pollution levels?

Controversies rage among the cognoscenti about whether to disrupt the tidy market-oriented accounts on behalf of values that cannot be priced. Many economists and statisticians dig in to resist change. "Even if past national income calculations are shown to be wrong," remarks World Bank economist Salah El Sarafy, these economists "still wish to maintain their continuity," valuing the security of the status quo above all. On the other hand, many of those who favor reform fear that with translation into money terms environmental assets will inevitably be undervalued. The choice may well be between monetary accounts that achieve precision but are increasingly detached from reality, and broader measurements of assets and costs that are less precise but tell us more about our world.

Heretofore, most national and international efforts to add environmental costs and benefits to national income tabulations—those of Norway and France are prominent examples—have taken the form of satellite accounts. Both France and Norway have established extended supplemental systems of resource accounting. In Norway accounts have been drawn up for raw materials like oil and minerals, for natural assets like forests and fisheries, and for environmental resources like air, water, and land. Except for natural resources, which are also valued monetarily, the Norwegian measures are represented only in physical terms. France's "national patrimony accounts," under development since 1971, follow much the same pattern, with most measures expressed in physical units and monetary values included for resources that play a role in the market-place.

An international group of economists spent seven years over-hauling the widely followed U.N. System of National Accounts. The revised SNA model, published in 1993, lays out extensive environmental satellite accounts in both quantitative and mon-etary terms. Environmental costs of economic activities, natural asset flows, and expenditures for environmental protection and enhancement are tabulated in flow accounts and balance sheets that maintain the traditional SNA categories. Applied to accounts of an unidentified country, the model dramatically demonstrates how figures for net domestic product (NDP)—and particularly for agriculture and mining—fall sharply when they are calculated in terms of "Environmentally Adjusted Net Domestic Product" or what the SNA authors call "Eco Domestic Product (EDP)." Under EDP-based values, the NDP is reduced from 25 percent to 7 percent, and in the mining and agricultural sectors from 73 percent to 5 percent and 20 percent to 4 percent, respectively. These unsettling calculations are consigned to the sidelines, how-ever, while changes in the central SNA accounts are less than earthshaking: In the revised accounts, for the first time natural assets (i.e., proven reserves of minerals, fish stocks, and forests) and changes in their size were included in asset balances, but no other changes were made.

Although this country is likely eventually to adopt the U.N. accounting revisions on an official level, changing Americans' calculation of value will require a much more profound seismic shift. The United States Bureau of Economic Analysis has lagged behind European countries and, to a lesser extent, Japan, on environmental accounting—and even if satellite accounts are publicized, it is unlikely that they will affect our economic think-ing much. As long as the GNP holds center stage, environmen-tally adjusted tabulations can be written off as one more bunch of depressing statistics, regardless of how dramatic they are. Be-cause the full scope of environmental value can never be cap-tured in the hard monetary terms we are accustomed to, the EDP inevitably seems less "real."

If we cannot adequately count the monetary cost entailed in the destruction of our life-support systems, or the benefit of maintaining them, the alternative is to deflate the importance of GNP relative to other calculations of value. Doing so, however, will cut deeply against the grain. The idea that the human economy is materially subordinate to the economy of the biosphere confounds our preconceptions about our own preeminent scale. We strenuously resist the idea that this modest-sized planet we now traverse at will can veto our ingenious improvements on the place.

Using the market to solve our waste problems, on the other hand, has a very congenial sound. What does not sit so well is what has to come first—using government policies to reshape the way the market works.

Admittedly, the line between regulation and using market incentives is sometimes hazy. One of the most significant regulatory innovations of the past few years—the "take-back" provisions by which the German government is mandating producers to assume responsibility for postconsumer waste—will profoundly affect the shape of the marketplace. As a result, German corporations and international enterprises with stakes in the German market are starting to compete in developing processes and products that conserve materials and minimize waste.

Even in the United States regulations have forced some market development. Procurement mandates and landfill bans have increased demand for recycled materials. Progress in spurring market incentives could come even more swiftly if all levels of government stepped up procurement of goods along guidelines of meeting strict environmental standards and incorporating recycled materials. Looking at opportunities, the Clinton mandate for recycled content in paper products is just the beginning. The federal government alone spends $360 billion to $480 billion, 6 percent to 8 percent of the GNP, purchasing goods from private-sector suppliers. This purchasing power could be redirected toward companies that incorporate strict environmental standards and products that foster recycling or reuse. For example, the General Services Administration could insist on higher than average fuel standards in its fleet of half a million vehicles.

Where an environmental goal can be achieved without the explicitly directing pressure of regulation, however, the market can do so more efficiently than regulations, often by making the polluter pay for waste disposal, remediation cleanup—or avoidance. Many of the most effective market-based mechanisms derive from the "polluter pays" principle or its corollary: Pollution avoidance pays. Thus costs formerly treated as "external" to market transactions are "internalized," or included in prices paid by producers and consumers. For example, charging consumers according to actual quantities of garbage clearly cuts household waste volumes: Generators must pay for what they dump. Similarly, bottle and can deposit-refund programs have proved most successful at pulling recyclables out of the waste stream. In the case of deposit refunds, consumers are paid—in effect by themselves—*not* to pollute.

Deposit-refund programs may offer the only feasible way to achieve certain environmental objectives. For example, as regulations tighten, the impetus to dump oil or chemicals down the sewer or in an empty lot increases. Particularly when oversight is thin, it is difficult to detect this sort of illegal dumping, and penalties for those who are caught too seldom serve as an effective deterrent. Incentives, on the other hand, appear to offer a promising alternative. Spare-time mechanics may well opt to turn in used motor oil for recycling or disposal if they will thereby recapture part of the purchase price. (Deposit-refund systems probably work best for wastes that maintain their original identity during use, like car batteries, dry-cleaner filter cartridges, pesticide containers, solvents, and motor oil.)

Emissions trading also offers a potentially powerful market-based tool for curbing pollution—relying on transferable rights to certain quantities of emissions. Trading permits generators that reduce pollution below permissible levels to sell emissions credits to firms that find it more difficult to comply with regulations. Thus trading can reduce costs of compliance to all and provide money incentives to cut emissions as much as possible. In 1992, 3M's manager of pollution prevention programs, Thomas Zosel, described the advantages of trading to his company:

> If new technology [for pollution control] doesn't perform as we expect it will, without trading, we may have to rip it out. With trading, we can keep it, improve on it while it is in operation, and

buy credits to make up the difference between the technology
and what the regulations require.

While it has been much discussed during the past decade, until
now trading has taken place largely within industrial installations
where, say, sparing application of solvents on circuit boards may
compensate for heavy flows in cleansing metal presses. It is only
beginning to be tested on any scale, most importantly in Southern California, as a result of endorsement in the Clean Air Act
of 1991.

Although authority for emissions trades is only implied in the
1987 amendments to the Clean Water Act, the EPA has encouraged trades aimed at meeting or bettering permitted pollutant
levels through transactions between point sources or between
point and nonpoint sources. In deals arranged between sewage
plants, for example, one plant would help finance the installation
of sophisticated aeration equipment in the other in order to
achieve higher joint reductions at lower cost. Another variant
would aim at inducing farmers to curb runoffs. Here, sewage
treatment plants might pay farmers to use management practices
that would cut down on applications of fertilizers or plow under
livestock wastes. This sort of buy-up has a parallel in the "cash
for clunkers" program initiated in 1992, which would encourage
federal utilities or industries to buy the right to continue polluting processes by reducing pollution from vehicles. "Stationary-
source" polluters could buy up old cars for scrap (the oldest,
presumably from the early 1970s, which emit perhaps twenty
times the pollutants of 1990s models, would fetch the highest
prices) and apply the credits for diminished pollution to their
own "accounts."

Auctioning off permits for use of agricultural chemicals—with
volumes authorized according to the absorptive capacity of a
given watershed—offers another way to cut farm-generated pollution through the workings of the market. In combination with
the bidding process, the relative scarcity of permits could effectively internalize costs of chemical wastes by raising users' expenditures in inverse proportion to the condition of surface and
groundwaters in their particular area.

When one considers the advantages of these market-oriented
mechanisms for curbing waste and pollution, it is important to

realize that their operation is anything but automatic. None are self-generating or self-sustaining. Despite the attractions of air pollution emissions trading as an idea, getting it to work well takes a lot of doing. The market must be competitive, that is, relatively unhindered either by overly complex regulatory provisions or special concessions to local interests, the transactions costs must be reasonable, the emissions must be measurable, and the rights offered must be clearly defined. Considerable "organizing" of the market by both public and private entities must precede—and then accompany—the functioning of these mechanisms. Although emissions trading has been tried much more extensively in this country than in Europe or Japan, awkwardly functioning markets for emissions rights have impeded progress. The extensive, often complicated procedures for external trades have raised transactions costs and introduced considerable uncertainty about property rights.

Once these mechanisms are in place, we will still need government oversight of environmental outcomes. Even if emissions trading takes off, a number of problems, such as the shifting of pollution to vulnerable regions, will remain to be ironed out. As emissions trading got underway in 1993, New Yorkers, worried about Midwestern pollution pouring into Adirondack forests and lakes, moved to block sales of emissions permits by the Long Island Lighting Company.

In the longer run, we must go further than mechanisms like deposit return or emissions trading will take us. The reality now is that singly and together, none of the market-based methods we've looked at will do enough. If we seriously intend to stanch the flow of waste, we will have to raise the prices of the materials supplying our production through broadly applied taxation. Only thus will we be able to adequately assess producers and consumers for the full costs of both pollution and of natural resources. Making the polluter pay is still the basic idea. Because it is well-nigh impossible to accurately apportion the responsibility or exact full reparations after the fact, however, we will have to start at the front end. We need to change our tax system to substitute levies on resources, pollution, and waste for many of the income and sales taxes that currently supply most public revenues.

How could we possibly sell materials taxes to ourselves? Before the election of the Clinton-Gore ticket, responses to various energy tax proposals in the United States had run into a solid wall of ideological resistance and special interest antagonism. The arguments against such taxes, that raising energy prices or cutting energy use would drain U.S. competitiveness, curb growth, and hurt the poorest most, were familiar and widely accepted. Between 1988 and 1993, even the Gulf War could not impel Congress to raise the gasoline tax by more than a nickel, or to seriously consider a carbon tax to cut CO_2 emissions.

Now, with administration support, prospects for taxing energy may be improving, but as the summary dismissal of the proposed Btu tax in 1993's budget agreement demonstrated, serious proposals for far-reaching environmental levies would undoubtedly arouse a firestorm of opposition. First of all, the bonds between Americans and their automobiles remain strong—undoubtedly reinforced by the lowest gasoline prices since the Arab oil embargo in 1973. Even after a multiyear economic downturn, in 1992 census data showed an increase in the number of households with two or more vehicles, and a decline in the number of people riding to work in car pools or taking mass transit. More and more Americans were moving to gas-guzzling light trucks as a substitute for the family car. Second, some industries would be big losers. These include the mining sector, bulk chemicals, power-plant construction, metal-smelting plants (especially aluminum from virgin materials), truck manufacturers, and cement producers. Perhaps anticipating a change of environmental course that failed to materialize, in October 1992 Dow Chemical's CEO Frank Popoff launched a shrewd preemptive strike, calling publicly for more efficient environmental controls in the form of a "pollution-added cost in product pricing," to be applied through industry efforts to break out environmental costs lumped into overhead. At the same time however, he explicitly opposed taxation, which is the most efficient way to add environmental costs.

If taxes on materials were understood, their use ought to have substantial appeal for the larger public. In our political context, that is of course a very big "if." Yet taxing materials instead of income would not unduly penalize the average taxpayer, and would spur economic growth. By taxing environmental "bads" rather than productive "goods," like labor and capital, we would be correcting the market distortion entailed in the "externalizing" of environmental costs. Thus we should increase overall efficiency.

Experts have devised a number of different environmental taxes, including levies on carbon content of fuels, on polluting emissions—the chief being carbon dioxide, carbon monoxide, sulfur dioxide, nitrous oxides, methane, and nonmethane volatile emissions—on toxic chemicals, on fertilizer and pesticide sales, on chlorofluorocarbon sales, on paper and paperboard made from virgin pulp, on products like batteries, and on groundwater depletion. Until now, except for a levy on the sale of CFCs, largely to prevent windfall profits, and the gas guzzler excise tax,* the United States has shied away from these sorts of charges or taxes. In Europe and Japan, on the other hand, environmental taxes are becoming quite common. France taxes the emissions of 870 industrial plants and the discharge of industrial waste into the sewage system; Germany also taxes water pollutants and hazardous waste; Norway taxes pesticides and fertilizers; Sweden has introduced an entire package of environmental taxes; Finland and the Netherlands are following suit. So far it is not possible to judge the effectiveness of the extant taxes: Except in the case of U.K. levies on leaded gasoline, charges have been too low to change behavior very much.

Comprehensive environmental taxes, however, would work very differently than any of the narrowly targeted levies now being tried. The idea would be to preserve "revenue neutrality" by using environmental charges not to finance specific environmental or other programs but to substitute for other revenues. While taxes on fuels, raw materials, especially toxics, water, and perhaps emissions, were being raised, other taxes would be proportion-

---

*Two other excises—on certain chemicals, to support the Superfund trust fund, and on petroleum, to support the Oil Spill Liability Trust Fund—are unrelated to the toxicity of products or services. They offer virtually no incentives to substitute less hazardous substances or to reduce wastes.

ately reduced. It must be emphatically stated that the overall fiscal burden on businesses and individuals would not increase. The tax revenues generated by materials taxes would be "recycled" to reduce other taxes. When the tax burden is shifted from capital and labor income, use of the increased income may be expected to raise productivity.

German economist Ernst von Weizsäcker explains more clearly than any other exponent of environmental taxes how comprehensive taxes could work. Looking at the central question—how high should taxes be set in order to reorient production and consumption?—he estimates that, according to the "polluter pays" principle, internalizing environmental costs in West Germany would require tax receipts amounting to about 10 percent of GNP. The sudden imposition of a tax that high, however, would ruin entire industries and entail very great adjustments for society as a whole. Thus a tax equivalent to external environmental costs should be phased in gradually, perhaps over twenty to thirty years at a rate of 5 percent to 7 percent a year.

Studies of consumer responses to relative prices of petroleum-based fuels demonstrate a strong correlation between higher prices and lower consumption. Petroleum and diesel fuels have been taxed very heavily in Japan and Italy, somewhat less so in other Western European countries, and hardly at all in the United States and Canada. Allowing for the effects both of population density and of per capita income, there is a remarkable consistency between price and consumption in Japan and Italy, where fuel consumption is about a quarter of that in the United States, and in the other European countries, where consumption is a bit less than half ours.

Although our energy-use pattern under the new regime would look quite different from the present one, we'd still have plenty of power and mobility. Considering the possible effects of a 5 percent to 7 percent annual price hike for petroleum in relation to observed fluctuations in the price of oil during the 1980s, von Weizsäcker predicts much more fuel-efficient cars and the availability of plant-based fuels, massive replanning of freight traffic, and modernization of shipping facilities and rail networks within ten years of levying taxes. The next period would see a drastic drop in petroleum consumption, and massive shifts of freight to rail and waterways.* With rising prices for petroleum would come a revival of transport powered by the sun, water, wind, and muscle, including high-tech sailing ships. The phasing out of petroleum for transport could be completed, at least in Europe and Japan, in about forty years.

Many arguments will be advanced against this profound disruption of the status quo. First and foremost come forecasts of economic decline. Proponents of this view assert that investments necessary for reducing resource consumption and waste will drain away funds otherwise available for expansion and innovation. However, current high taxes on human labor also result in considerable investments to avoid labor costs. Reducing taxes on labor, and thus costs of labor, should spur employment, thereby releasing public funds, even some now allotted to welfare.

By significantly improving natural-resource efficiency, environmental tax reform would also free financial resources for investment in expansion and innovation. A 1992 EPA study came to the conclusion that a carbon tax aimed at reducing CO_2 emissions would cost the economy $2.7 trillion over the next forty years, but would reduce energy bills by $5 trillion, for a net saving of $2.3 trillion. Savings on pollution control and health care should further swell net disposable wealth. Cuts in existing taxes on business profits and value added would give enterprises more room for maneuver. Moreover, new demand for energy-, water-, and material-saving products should also result in a new wave of invention and entrepreneurship.

Though our national experience conditions us to believe that growth must go hand in hand with rising energy consumption, nowadays that correlation clearly does not hold. As von Weizsäcker points out, high energy prices appear to have stimulated both efficiency and growth in various OECD countries. Between 1975 and 1991 those maintaining high energy prices at home, Japan first and foremost, but also the European Community, did discernibly better (in terms of GNP per capita, the external trade balance, the stock market index, and full employment) than the cheap energy economies, i.e., the United States and also the USSR. In Japan, overall energy use declined by 6 percent be-

---

*While trucking boomed in the United States during the 1960s to 1990s, in countries with higher fuel prices railroads and shipping fared much better.

tween 1973 and 1985, while per capita GNP rose 46 percent.

Once nations embraced them fully, comprehensive environmental taxes should also measurably improve the atmosphere for investment. An environmental levy running at a predictable pace over three or four decades would establish a climate of stability for business decisions about the environment that is sorely needed. Compared with the chaotic profusion of environmental legislation and fluctuating markets for energy and materials that we might expect otherwise, the certainty of the taxes would foster longer-term investment strategies. Taxes would provide a continuous incentive for cleaner technologies. Moreover, when the tax reform showed its effectiveness in curbing pollution and conserving resources, we would be able to start clearing away the regulatory underbrush.

Environmental taxes will not be a panacea. Like all other policies, they will have unintended consequences, frequently arising from the ingenuity of those who would evade their reach. Once these taxes were on the books, battalions of lawyers and accountants would go into high gear, figuring how to make them work to the advantage of their clients in ways their advocates had never foreseen. We would still need the visible hand of government to set standards, lead the way with public education and technical development, monitor environmental outcomes, and oversee enforcement. For example, taxes on emissions, requiring extensive monitoring, would clearly necessitate a lot more hands-on government involvement than those on raw materials, which could be levied at the point of sale (though raw-materials taxes should have a real effect on curbing emissions as well). Where there is acute danger to health or habitat, we will continue to require speedy and direct measures in the form of regulations.

We will also have to fine-tune both domestic social consequences and foreign trade impacts. At least in the short term, materials taxes are clearly regressive. Because the poorest would suffer most from higher prices, redressing inequities would necessitate a package of compensation measures. In addition, where certain regions are hit particularly hard, short-term tax offsets would be called for. Internationally, where the adoption of environmental taxes by particular countries created short-term trade advantages for their rivals, we would have to devise appropriate policies in GATT and other forums. (Taxes based on raw material content would create particular administrative problems regarding imports.)

Working as they do through the pricing mechanism, however, the taxes would eliminate a number of administrative problems we now face. Unlike most current regulations, taxes would not deal separately with air, water, and land, pushing pollution back and forth across media. Instead they would curb materials flowing into all three, often at the point of sale. Because it is far easier to monitor charges on sales than to enforce compliance with any sorts of pollution controls we have yet devised (such as state-of-the-art control technologies, emissions levels, units of toxicity, flows of materials, degree of contamination, or effectiveness of cleanup), taxes can keep administrative costs, both of business and government, to a minimum. The necessity for judicial oversight diminishes, and with it enormous expenditures on legal jousting and litigation.

Materials taxes also allow us to sidestep many of the intractable scientific questions about relative risks and benefits, which in the absence of massive additional analysis we are unlikely to be able to answer with any certainty in the near future. Although we may never fully understand how those many thousands of chemicals are affecting us and our habitat, or in which instance reduction, reuse, recycling, or substitution would make more economic and environmental sense, increasing parsimony will greatly lessen the need to know. In the current Superfund program, for example, we cannot avoid complex choices about risks and responsibility while we are cleaning up the results of past excesses. Indeed, we will always need to assess lurking dangers and evaluate viable substitutes. In the future, however, we would be much better off finessing as many of those tough questions as possible by severely constricting sources of waste and pollution.

Though current strategies for waste reduction have brought significant progress, they fall far short of what is needed. Once the first easy victories are over, ongoing efforts to spur pollution prevention by industry seem prone to slide toward the kinds of command-and-control regulation that have fallen short in effecting end-of-pipe controls. Environmental taxes, however, would abate the need for micromanagement of technologies and systems by environmental officials. Again, taxes internalizing environmental costs would lessen the need to mandate choices among the alternatives of reduction, reuse, and recycling. Whatever works best, pricing resources at their full environmental cost would make all the conservationist alternatives cost-effective.

Once undertaken, the taxes would require far longer to work in the United States than in other countries. The synergy between our vast landscape, our autos, and our highways has allowed Americans to claim an unparalleled expanse of living space. Our scattered population has ensconced itself in separate one-family dwellings and in far-flung suburbs to an extent unheard of, and impossible, in most other places. Whether the distribution of people in space might change or new communication and transport technologies might emerge to accommodate resource parsimony is impossible to foresee.

Ideology will also increase our resistance to this new sort of governmental fiscal intervention and its effects on our mobility and consumption longer than anyone else's. After the major tax reform of the mid-1980s, the idea of disrupting the whole structure again to push conservation may seem excessive to many Americans. In her confirmation hearings as deputy director of OMB (Office of Management and Budget) in January 1993, Alice Rivlin voiced this wariness—even regarding what was to her the highly appealing cause of taxing consumption rather than savings—saying that "the complexity and disruption" of changing the tax system was probably not worth the effort.

In the near term, therefore, prospects for acceptance of comprehensive environmental taxes are much more promising in Europe, and perhaps Japan, than in the United States. Clearly, however, we need this potent and streamlined market mechanism more urgently than any of the countries that are likely to adopt it ahead of us. For the United States, comprehensive environmental taxes would represent a very bold step. Yet they could at one blow cut away large swaths of the bureaucratic thicket that always dog American regulatory efforts, increase both environmental and economic efficiency, and goad U.S. business into new and productive directions. They could, as it were, free the American genius for solving problems.

Instituting a system of environmental taxes, while a major undertaking, nevertheless represents our best shot at expanding
our progress and prosperity into the twenty-first century. To
work, the taxes would have to be undertaken for the long term,
with phased increases reliably scheduled. Such a program would
require a consensus transcending the abrupt shifts in policy usually attendant upon American partisan politics. Unlikely as such
a social compact may seem, achieving it is essential to facing both
the environmental and economic realities of the next century. It
may be necessary for continued human life on this planet.

13
~~~~~~~~~~~~~
Progress and Plenty in the Twenty-first Century

We have not inherited the world from our forebears—we have borrowed it from our children.

—KASHMIRI PROVERB

Regrettably, democratic societies rarely respond until a crisis has erupted.

—JAMES E. SCHLESINGER, *The New York Times*
(January 4, 1989)

PORTRAYING THE REAL dimensions of the environmental crisis we have created is a risky business. The answer to the question of whether the glass is half empty or half full must be half empty, at best. Because "negativism" about our system is un-American, however, anyone saying so is in danger of losing his or her audience. The individuals and groups now working to stir up a backlash against environmentalism play on this response. Doomsayers are tiresome, they say, and at their worst may be plotting a Big Brotherly assault on the economic freedom that has made our country great.

In fact, as this book has argued, the urgency of the crisis requires solutions that play to our national strengths. We do not have time to overhaul our political culture before attacking the problem of waste. Thus the compelling need for the approach outlined here: Our best shot at reversing the trend toward environmental overload inherent in this century's burst of material growth is to use our favorite tool, the market.

In order to engineer the profound redirection of market activity that will salvage our material world, however, we need to keep the gravity of our situation in sharp focus. As a people, we have never been willing to purposefully change our system very much except when driven to do so by extreme need. We have noted the main instances when crisis forced change in American political behavior. In the 1890s, growing physical chaos in our cities propelled us to create new municipal institutions to organize urban life and provide such basics as streets, lights, and garbage disposal. During the Great Depression, so many people were jobless, homeless, even hungry, that the whole society assented to the governmental programs bolstering social welfare, despite their diminution of our treasured individualism.

Although our present debacle dwarfs those we faced before, the gradual nature of environmental deterioration makes it difficult to grasp. We know that population and development are increasing now at an exponential rate, but what does this mean? Donella Meadows, coauthor of *The Limits to Growth*, represents for many the Cassandra-like voice they would rather not hear, warning that the party's over. Indeed, her image of the pond being choked by lily plants drives home forcibly the insidious impact of growth on the physical environment. In a simple riddle for children, Meadows illustrates "the suddenness with which exponential growth reaches a fixed limit."

> Suppose you own a pond on which a water lily is growing. The lily plant doubles in size each day. If the lily were allowed to grow unchecked, it would completely cover the pond in thirty days, choking off the other forms of life in the water. For a long time the lily plant seems small and so you decide not to worry about cutting it back until it covers half the pond. On what day will that be? On the twenty-ninth day, of course. You have one day to save your pond.

As numbers of people increase exponentially, the human race is in a position analogous to that of the pond's owner: We will not know what we're losing until it's gone. From 1790, when there were slightly less than 1 billion people on earth, it took two hundred years to reach the 5.3 billion population of today.

That total may nearly double in the next thirty years. The effects of all those people and their quest for sustenance and prosperity on our common habitat is for us largely unimaginable. Nigeria, for example, which has been importing food to sustain its one hundred million people, may see its population jump to nearly three hundred million in that period. China is making the most determined effort of any nation on earth to control its number of births, but by 2025, even if the nation's population police have succeeded in holding the current 1.2 billion Chinese to a "moderate" increase to 1.6 billion, surging national economic growth will still pose enormous problems for China and its neighbors. While the automobiles owned by 1 percent of Chinese now jam the country's roads, millions more entrepreneurs are determined to share the advantages of private mobility. In a country where air pollution already threatens respiration, the pressures of people on the environment could be almost literally earth-shaking.

Our own specific problems, though different from those of the Asian and African giants, are also directly tied to theirs. Numbers of people in the United States will probably rise fairly modestly, from about 256 million now to perhaps 335 million in the next thirty years. Nonetheless, as a result of our own and others' economic development, we will have much to deal with—both the stresses created by our accumulating physical pressures on our own living space and those stemming from the effects of other countries' escalating population and development on our shared habitat.

If current trends continue, air pollution from burgeoning manufacturing industries and growing auto traffic can be expected to diffuse ever more widely. Less obvious but more irreparable to our posterity, the leveling of tropical forests and pulverizing of coral reefs will extinguish countless plant and animal species, perhaps a majority of those now in existence if the present pace of destruction continues for the next fifty years. Worldwide grain harvests, which rose by 40 percent per person between 1950 and 1984, are now falling by about 1 percent a year. As we have seen, the world seafood catch, which climbed from twenty-two million tons in 1950 to one hundred million in 1989 has also been declining since then: in an effort to stave off further declines the U.S. Commerce Department in 1993 planned to reduce fishing of some species by 50 percent. From now into the foreseeable future, as Paul Kennedy points out tellingly in *Preparing for the Twenty-first Century,* the United States and all other advanced industrial nations will be forced to face very directly the demographic dilemmas of their near and far neighbors. To put it bluntly, from now on would-be immigrants can be expected to be scrambling aboard through every open porthole.

In addition, the development pattern of our own highly mechanized society presents us with many difficult choices. In America today development pushes against the environment everywhere. New York City proposes to pay $272 million to sequester reservoir waters from the runoffs of new and proposed housing tracts, as an alternative to spending $8 billion for filtration plants. In the Mississippi floodplain, human settlements and the towering levees erected to protect them fall again as the great river reestablishes its domain. Nationally we squabble over owls versus loggers while rows of planted trees replace the flora and fauna that evolved over many millennia in the old-growth forests. On well-trampled Western prairie, succulent grasses and plants give way to sawgrass and thorny thistle. Canada shuts down codfishing off Newfoundland, and in Washington, the U.S. Department of the Interior buys up salmon-fishing rights from Greenland in an effort to restore salmon to severely depleted New England rivers and streams. Ornithologists debate the reasons for the precipitate worldwide decline of songbird species, and biologists note with mystification and alarm the decline and disappearance of myriad frog species from ponds in every latitude and longitude. Like the owner of the lily pond, we do not notice the perilously diminishing numbers of most plant and animal species until they are very close to extinction.

As limits tighten, conflict and disorder are likely to increase. In the Middle East, battles over water may soon overshadow traditional sectarian warfare. In this country, in courthouses and through local elections, the cities and farms of the West fight over dwindling river flows and aquifers. As property owners join forces to resist meddlesome environmental restrictions on their right to pipe sewage or to level habitats on their own land, general anger mounts. In an embattled atmosphere, the option of illegal dumping and despoiling is likely to become more acceptable to some. If the social compact frays, the power of public pressure inevitably diminishes and the need to obey laws that can be enforced only with difficulty seems more tenuous.

In current environmental enforcement efforts, financial penalties have done the most to catch the attention of potential polluters. Yet with some reason stakeholders on both sides of the various environmental divides complain even now that we are spending outsize portions of our common resources without first determining priorities. Superfund provides the main example of lavish spending to accommodate stringent standards that rarely apply in other areas. The benefits to the health of humans or other species of, say, excavating and incinerating all the soil on several acres at a cost averaging around $37 million are certainly debatable. Clearly, we sorely need money elsewhere—for example, for sewage treatment, preservation of habitats, and research on technologies for reducing waste. Nonetheless, significantly loosening the regulations governing Superfund sites, perhaps by broadening options for sequestration of toxics rather than cleanup, may well encourage more evasion of existing rules about disposal and emission of toxics.

As in health care, environmental triage is likely to present major political difficulties. Even more than with health care, though, much of the scientific data needed to ground decisions about regulatory priorities is not available. For individuals and communities as for government, setting priorities for action amid all the uncertainties dogging the new fields of environmental science and public policy is a recipe for confusion and frustration. Sometimes, as in Rye Brook, New York's experience with an EPA cleanup of nine backyards contaminated by mercury starting in 1991, it seems as if you're damned if you do and damned if you don't. Two years and $4.5 million later, the local residents appreciated EPA concern for their health, but greatly regretted the agency's approach in bulldozing their yards, felling trees, and tearing up lawns.

In this new world of inexorably tightening limits, the frustrations can only be expected to increase until we shift course 180 degrees to make regulation much less central to good behavior. As we have seen, neither "good science" nor tighter regulation nor entrepreneurship—nor all of them together—can salvage our living space unless we realign our economy to make conservation the bedrock of the good life. As this book argues, we can bypass many of the seeming contradictions between economic progress and environmental quality by redirecting market incentives to found our economic welfare upon conservation rather than exploitation of materials.

As this book was being written, the early record of the Clinton administration demonstrated just how hard the necessary changes are likely to come. Symptomatic was the decimation by Congress of the modest proposed increase in the national tax on energy, and the administration's difficulty in imposing even token fees to bring exploitation of government lands in line with market values. The $17 billion initiative for government–industry cooperation in nurturing new technologies announced soon after the inauguration of President Clinton also swiftly fell victim to congressional outrage over "massive government spending." Back at the grass roots, carpooling efforts set in motion during the energy crisis of the late seventies had lost so much momentum that state governments in polluted areas like Connecticut were threatening to fine companies to force cuts in the commuting time of their employees.

While policy can ease economic adjustments, moving from exploitation to conservation would entail an enormous change in the way we think about the material world. Getting from here to there through a revamped market would inevitably bring a far-reaching reorientation of attitudes. Even if we choose this path we will have to become accustomed to viewing the pursuit of happiness in some novel ways. To achieve the same degree of well-being while using far fewer materials, we will have to adapt our idea of progress to an era in which improvements in consumption do not automatically equate with increases in production. Very simply, we must begin to judge our well-being by the quality and quantity of what we *use* rather than what we *own*.

For Americans this change need not entail any real loss of individual freedom, but the material basis of our individualism must certainly be greatly altered. As we make the transition, the grounding of our freedom in secure control over property, a touchstone for Americans since our earliest forebears, will be transmuted in important ways. If we are to survive with habitat intact, the new economics of daily life will carve deep inroads into the self-sufficiency of our physical lives. This does not mean that the way we live will be less comfortable but that we will base our sense of individuality much less than previously on our ability to accumulate (and dispose of) things. As environmental engineer and philosopher Amory Lovins puts it, "I've never been interested in just doing with less. I'm interested in doing *more* with less." Though a person's home may still be his or her castle, the importance of inalienable title to physical possessions will diminish relative to that of securing access to both objects and information.

A few signs of this new way of thinking about progress and plenty are evident now. In the spring of 1993, offices of the Patagonia clothing company announced their intention to limit growth in the United States and eventually halt growth everywhere. In the future, they declared, given the state of the environment, they will be designing clothes that people will not need to change so often. In their words, "the future of clothing will lie in a few good clothes that will last a long time." In the summer of 1993, a new species of cooperation among the Agriculture Department, the Environmental Protection Agency, and the Food and Drug Administration brought a compact to jointly curb the use of pesticides. The agreement among the three sometimes warring federal agencies signaled a profound change in culture at the level of the federal government.

At around the same time, the government of California, like those of other Western states, broke with over a century of precedent by diverting water from notably wasteful agricultural uses: New rules required farmers to pay much more for the scarce resource and returned large flows to rivers and streams to protect fish and repair damage to wetlands. Looking across the Atlantic, in the spring of that same year the Dutch government changed its country's history in truly radical fashion by letting the sea flow back into an expanse of farmland reclaimed by dikes a century ago—ending a costly contest with nature that was felt in sinking water tables and polluted fields and aquifers. Similarly, in Israel, engineers restored the Jordan River to its original course, turning dessicated farmland, reclaimed with more fervor than under-standing during the 1950s, back to marsh. And in China, while urban entrepreneurs were busy discovering capitalism, scientists were reviving traditional methods for increasing yields while cut-ting fertilizer use in heavily cultivated Kiangsu Province, by cy-cling wastes among fields, pigs, and fish farming.

If and when these separate and tentative sorts of progress cul-minate in a full-fledged about-face, how might our lives be changed? Reducing our consumption and disposal of materials by relying on *use* rather than *ownership* to supply our physical needs and desires would fundamentally change our lives. In ac-commodating the idea of using things rather than owning them, we would be entering a wholly new era of production and con-sumption.

Taking computers as an example, the hulks of outdated main-frames and personal computers have become a sizable disposal problem. In the United States we will discard around 150 million machines during the next decade, paying perhaps a billion dol-lars for disposal and using up landfill space equivalent to stacking up fifteen Empire State Buildings. While computer manufactur-ers here and in Europe are exploring ways of dismantling and recycling both metal and plastic components, in the future, longer-term advantage lies in refurbishing and reuse. First of all, however, manufacturers have to design machines whose com-ponents can be interchanged easily, not only to make repairs but also to incorporate newer systems.

As new patterns evolve, the move toward leasing should be driven by demand as well as supply. For in many cases leasing should be more convenient for consumers. Instead of buying a new computer, lugging the old one off to be dumped, and get-ting the new setup established—with appropriate adjustments in the work station and printer as well as software installations—you would merely have to lease a desired level of service from, say, IBM or Microsoft. After that, the manufacturer/lessor would in-stall the appropriate hardware, software, and work station, service it when necessary, and update components when new technolo-gies or software came on line, while instructing lessees in the latest uses of their instruments. Service contracts would of course be renegotiated at given intervals, and whenever consumers/ users decided to change levels of function. Thus competition between manufacturers/lessors would be keen, and innovation in hardware and software as well as in design for reuse and re-cycling could be expected to remain high.

It takes a greater leap to imagine most Americans giving up ownership of their beloved automobiles. Many Americans do of course lease cars now, opting for a convenient financial arrange-ment and often favorable tax benefits. American car manufac-turers are also currently pushing leasing operations as a way to spur sales of new cars through finance companies that arrange long-term rentals. All parties assume that after several years the cars will be sold as used, ending the connection between pro-ducer and product. In the economy of use, however, the lease would become the ultimate end product, and the car the means for delivering it. Manufacturers would retain ownership from cra-dle to grave, and rather than working to unload aging merchan-dise, they would work to maintain and upgrade it.

While the auto industry may presumably be loath to make such an institutional change without external pressure, consumers should not have to be coerced: The new system would offer great convenience and diminished risk. They could sign up for use of an automobile in several different places where they might ex-pect to need one—say, in the town where they lived, in the city or cities where they traveled on business, in their parents' home-town or a vacation spot. When traveling they could step off one or another mode of high-speed mass transit and into a car, wait-ing at station or airport, all paid for by their regular monthly fee. Again they could buy various levels of service, taking into account different sorts of functions, and presumably extending just about everywhere in the world. Service and repair with, of course, a replacement car for the repair interim, would come with the deal.

Goading businesses into marketing strategies based on dura-bility rather than the latest model change will require reshaping incentives both by raising the cost of materials and by making producers responsible for disposal of their products. Even more than vendors of electronic durables, auto manufacturers proba-bly need to be pushed to take the sort of long-term responsibility for their inherently degradable product that the "use" system would imply. In addition to materials taxes, "take-back" provisions like those already enacted in Germany will probably be needed to force a full-scale shift. The two in tandem will fully internalize the environmental costs of production. Thus they will heighten the advantages of changing product and process designs to conserve materials and avoid their diffusion in waste.

Correspondingly, raising the cost of materials and requiring manufacturers to assume responsibility for disposal should drive home to them the advantages of leasing. Enabling manufacturers to retain control over materials during both production and consumption cycles, the leasing approach greatly increases incentives for durability, reuse, and recycling, which yield big savings on both procurement and disposal of materials.

Where current market advantages are becoming evident, early manifestations of the economy of use are emerging. Since 1992, Dow Chemical has been providing customers who are phasing out ozone-depleting solvents with complete cleaning systems guaranteed to perform according to given specifications. They also offer a variety of support services and assistance with upgrading processes as technology and chemistry evolve. At IBM, product managers now design for the whole lifespan of a product, including "end of life" costs among other criteria. As a result, its Personal System/2 models 40 and 57 computers incorporate a new "snap" technology (instead of fasteners) to facilitate disassembly, and use only one plastic resin to foster recycling. The company also voluntarily takes back its products for disposal in Germany, the United Kingdom, Switzerland, and Austria. Like IBM, for the past several years Xerox has been designing products and processes to facilitate refurbishing, reuse, and recycling. Its first environmental design to reach the market was a customer-replaceable cartridge, containing the main elements for its smaller copiers, whose parts and materials are totally reusable and recyclable.

Further, in the energy sector, the "demand-side management" approach to electricity sales pioneered by California's Pacific Gas and Electric utility aims at fulfilling customers' changing needs by generating power more efficiently rather than by increasing sales of wattage. Currently, a large portion of PG&E's

effort is spent working with customers and users to analyze energy use and install equipment that conserves it.

In a different arena, preserving natural habitats for tourists, rather than exploiting forests for timber, land for minerals, rivers and streams for irrigation, or prairies for meat production, also encourages use rather than appropriation and consumption. Because tourism can be sustained indefinitely, generating jobs and income in the long term, it represents a much better investment than the alternative of materials extraction which will end when old-growth mining slag lies heaped across mountainsides, forests are fields of stumps, and soil erosion has made of rolling plains a landscape of dirt and sawgrass. Again, looking at the way we go through housing stock, enormous savings—with potentially high aesthetic payoffs—would result from revising tax and subsidy policies to favor renovation rather than construction of new housing tracts.

Without pressure, we are unlikely to move away from the economy of consumption anytime soon. Yet, in fact, a move toward leasing and use would push us farther and faster in the direction we are already going, toward an economy where the comparative advantage lies in services. Although the current shift is not evident in our output of waste, the U.S. manufacturing sector has been declining relative to services since mid-century. In 1992, service industries accounted for 72 percent of the U.S. GDP and employed 76 percent of workers. Combined exports of legal, business, health, and recreation services contributed most to overall U.S. export growth and amounted to around 30 percent of export totals in 1992.

Moreover, in our increasingly complex and vulnerable economies, the functioning of systems and quality of products depends more and more on service. As European economists Orio Giarini and Walter R. Stahel have pointed out, present-day economies of scale, interdependent technologies, and the flow of wastes from complicated technologies and swelling production have raised risks dramatically. Thus it is incumbent upon consumers at all levels to look much more warily not just at products but at the likely results of their utilization *over time*.

Today's key economic problems concern systems rather than commodities—how to deliver health care, education, and housing. And in buying products, we have to evaluate not only their design, but how we will learn to use them, what's involved in maintaining them, and what risks we incur from possible malfunctions or emissions. For a product like the computer, tools for learning how to use it, i.e., "software," may surpass "hardware" in economic importance. Similarly, with many electronic goods—such as copiers, printers, and video-conferencing equipment—the sale of a maintenance contract is becoming the norm.

If we acknowledge that the results of systems and the use of products over time are paramount in today's economic world, as Giarini and Stahel argue, standards based on design criteria need to be replaced by those evaluating performance. Technologies for on-site monitoring of changes in quality of performance, as, say, in railway tracks and bridges, cars and planes, or incinerators and sewage plants, should play an increasingly critical role in fostering long-term utilization. This book has shown that designing for long-term use and reuse of products and materials is eminently feasible. Making design for waste avoidance a key performance criterion, however, will require overhauling both systems and attitudes.

If we succeed in changing the main orientation of our material lives from ownership to use, we will also have to consider what services such as repair, refurbishing, and maintenance are worth in relation to manufacturing. Right now, prices of replacement parts for mass-produced goods are prohibitive—reassembled parts for a car may cost more than four times its sticker price. This discrepancy reflects not actual costs but marketing strategies emphasizing obsolescence and replacement. On the other hand, the value of skilled "manual labor" doing repairs is usually low. As the economic payoff of conservation rises, so should that of the services that make it possible.

Not incidentally, such a system should also boost employment. In facilitating comfort and convenience through service, leasing, and use rather than sale and disposal, human ingenuity and time will often substitute for the materials and machinery central to manufacturing. While it may prove most efficient for large manufacturers to superintend the reuse and recycling of many products, much of the inevitably labor-intensive repair work can best be carried out in relatively small workshops located conveniently for consumers/users. When materials become more expensive than labor, full employment may seem, once again, like a real-world possibility for a competitive U.S. economy.

More flexible access to the necessities and comforts of daily existence also fits well with evolving styles of family life. Leasing and servicing particularly serves the changing needs of women. The idea prevailing since the days of *Mrs. Consumer* that procuring and maintaining her family's possessions is a full-time job for wife and mother is quite distant from the current realities of the two-career family. Housewives today, though working toward parity as breadwinners, are still chiefly responsible for making the American home run smoothly. When the family group's general level of comfort and care is determined more by access to services than by the maintenance of the possessions themselves, much the housewife's custodial onus should disappear. When spouses are divvying up responsibilities for arranging services rather than performing domestic chores, it should prove easier to share household tasks between them.

While America grapples with its own specific set of problems, how we respond to the environmental crisis will also greatly affect what happens in the rest of the world. In aspiring industrial powers like China and in poor and densely populated Third World nations like Nigeria, their people's determination to replicate our wasteful and polluting industrial development patterns has very serious implications for their quality of life and for ours as well. If, instead, they can build their economic revolutions on conservationist technologies geared to twenty-first-century materials efficiency, we will all be much better off. But they will have a chance to take a different route only if we do so first, for our technologies will provide their leg up. Unless the United States and its fellow industrialized nations move posthaste to create the machinery and systems for a conservationist alternative, it may not be possible for the rest of the world to emulate us without making the same mistakes.

Developing the technologies for a conservationist mode of progress would not only smooth our own path, but would also give us leverage over our common global environment that we now sorely lack. Yet the technologies we need will be long in coming, even here, unless we purposefully redirect market incentives to achieve them. Can we do that? If we would salvage our living space and pass it on, alive and beautiful, to our posterity, we really have no other choice.

Notes

In bringing together all the separate elements that make up the story of our waste, I have been fortunate in being able to draw on many excellent studies in a variety of different fields. An account of my debts is found in the notes on individual chapters, which follow. I hope these sources will be as instructive to readers who wish to push further as they have been to me.

In citing statistical data, I have not attempted to maintain consistency in derivation of sources, but have incorporated figures from a wide variety of reports, indexes, and compilations, many of them published by various departments of the federal government. The latter are identified by their initials, principally the United States Congress's Office of Technology Assessment (OTA); the Environmental Protection Agency (EPA); the General Accounting Office (GAO); and the Office of Management and Budget (OMB). In any case, available statistics define the extent and impact of waste very imperfectly. While municipal collections provide information on consumption-generated wastes, we know much less about the fallout from production, which is handled privately. And the impact of these wastes in emissions and pollution are very hard to count or measure reliably. It is important to recognize this reality. Our imperfect knowledge of the world of waste is part of the larger problem treated at length in this book: the great gap between our enormous impact on the environment and our as yet limited understanding of that impact.

Works frequently cited have been identified by the following abbreviations and short titles:

NYT	The New York Times
WSJ	The Wall Street Journal
WP	The Washington Post
WW	WorldWatch
BAS	Bulletin of the Atomic Scientists
PNAAS	Proceedings of the National Academy of Arts and Sciences
PPR	Pollution Prevention Review
Democracy	Alexis de Tocqueville, Democracy in America (New York: Random House, 1945).
Earth Transformed	B. L. Turner II, ed., The Earth Transformed by Human Action: Global and Regional Changes in the Biosphere over the Past 300 Years (Cambridge: Cambridge University Press, 1990).
Garbage in the Cities	Martin V. Melosi, Garbage in the Cities: Refuse, Reform, and the Environment, 1880–1980 (College Station, Tex.: Texas A&M University Press, 1981).
More Work	Ruth Schwartz Cowan, More Work for Mother (New York: Basic Books, 1983).
Trash to Cash	Susan Williams, Trash to Cash (Washington, D.C.: Investor Responsibility Research Center, 1991).
War on Waste	Louis Brumberg and Robert Gottlieb, War on Waste: Can America Win Its Battle with Garbage? (Washington, D.C.: Island Press, 1989).

Chapter 1: What's the Problem?

As we reflect on how our prodigious economic drive has weighed on our physical environment, it is instructive and inspiring to hark back to the seminal work on this subject, considered by some to be the most important wellspring for the contemporary environmental movement, Rachel Carson's *Silent Spring,* originally published in 1962 by the Houghton Mifflin Company in Boston.

The growing interest in garbage and the waste problem more generally is signaled in several current trade books on the subject, notable among them William Rathje and Cullen Murphy, *Rubbish!* (New York, HarperCollins, 1992), and Judd A. Alexander, *In Defense of Garbage* (Westport, Conn.: Praeger, 1993). Several interesting studies of the garbage crisis also appeared earlier, in 1989, after the travels of the garbage barge *Mobro* and other symptoms of profound systemic overload had caught public attention: Notable among these are *War on Waste,* and *Rush to Burn: Solving America's Garbage Crisis?* (Washington, D.C.: Island Press), by the staff of *Newsday,* a daily newspaper published on Long Island. See also Allen Hershkowitz, "How Garbage Could Meet Its Maker," *The Atlantic Monthly,* June 1993, pp. 108–109.

Page 26, par. 4: Gregg Easterbrook's argument is made in "Indestructible Earth," *ECO,* June 1993, pp. 32–34. On *Garbage,* see "Disposable Diapers: They're Okay and You're OK." *Garbage,* October/November 1993.

Pages 27–30: For further elucidation of the issues related to human impacts, see the major source, the pathbreaking compendium *Earth Transformed* (see note, page 294), an authoritative and extraordinarily comprehensive account of the accelerated diffusion of materials within the natural environment as a result of recent human production and consumption and their impacts on land, air, and water. Chapter 1, "The Great Transformation," by Robert W. Kates, B. L. Turner II, and William C. Clark provides an excellent overview of a very large canvas.

Page 28, par. 4: On the growth in industrial capacity, see *Earth Transformed,* pp. ix, 158–177, 7–8, 220–309, 438–447.

Pages 27–28: On impacts before industrialization, see Ibid., p. 10.

Page 29, par. 2: On emissions generally, see Ibid., p. 6. On carbon mobilization, see pp. 395–406, and particularly p. 405. On water use, see pp. 12, 235–267, 277–292. On the increased use of synthetic chemicals, see p. 256. On pollution of the oceans, see pp. ix, 330, 313–329.

Page 29, par. 4: On riverine pollution, see Ibid., p. 248, fig. 14.16(b). See also Paul Kennedy, *Preparing for the Twenty-first Century* (New York: Random House, 1993), pp. 21–46 and 95–121.

Page 30, par. 3: The work of Robert U. Ayres has broken new ground on the conversion of materials to waste in viewing systematically the use of materials before, during, and after industrial processing. See Ayres's "Industrial Metabolism" in Jesse H. Ausubel and Hedy E. Slavovich, eds., *Technology and Environment* (Washington, D.C.: National

Academy of Engineering/National Academy Press, 1989), p. 26.

Pages 30–32: The anecdotal history of garbage disposal methods of different civilizations draws on a number of accounts, principally the seminal work of historian Martin Melosi, *Garbage in the Cities* (see page 294), pp. 4–8; Lewis Mumford, *The City in History: Its Origins, Its Transformations and Its Prospects* (New York: Harcourt Brace Jovanovich, 1961), pp. 73ff.; *Rubbish!*, pp. 32–40; and Alexander, *In Defense of Garbage*, pp. 1–3.

Page 33: The account of Vicon's aborted incinerator project is derived from *War on Waste*, pp. 123–125.

Page 34, par. 2: The discussion of cross-media exchanges is derived from *Earth Transformed*, particularly pp. 204, 317, 448.

Pages 35–37: The personal anecdotes about Lois Gibbs's crusade are drawn from her autobiography: Lois Marie Gibbs, as told to Murray Levine, *Love Canal: My Story* (Albany: State University of New York Press, 1982).

Chapter 2: Freedom and Wealth in the Minds of Americans

Page 43: Alexis de Tocqueville, *Democracy* vol. 1, p. 198.

Page 45, par. 2: on Americans' innocence of feudalism, see Louis Hartz, *The Liberal Tradition in America* (New York: Harcourt Brace and Company, 1955), pp. 233–237.

Pages 46–47: Regarding Americans' profligate use of land, see Daniel J. Boorstin's seminal work *The Americans: The National Experience* (London: Weidenfeld & Nicolson, 1966), pp. 93, 243.

Page 47: Francis J. Grund is quoted in Thomas J. Cochran, *Challenge to American Values: Society, Business and Religion* (New York: Oxford University Press, 1985), p. 26.

Page 48, par. 2 and 3: regarding Social Darwinism, John Kenneth Galbraith discusses the relationship between the American embrace of Spencer and the fact that the downside of industrialization had yet to become fully apparent here in *The Affluent Society* (Boston: Houghton Mifflin, 1958), p. 56. Galbraith discusses the rise of Social Darwinism in the U.S. at the same time as the rise of the great fortunes in Ibid., p. 59. On the American view of voting as analogous to consumption, see Ibid., p. 260.

Page 49, par. 2: On American ingenuity, see Boorstin, *The Americans*, especially pp. 26, 27; see also Cochran, *Challenge*, and David Freeman Hawke, *Nuts and Bolts of the Past: A History of American Technology 1776–1890* (New York: Harper & Row, 1988).

Page 49, par. 3: On the comparison between entrepreneurship in England and America, see Joel Mokyr, *The Lever of Riches: Technology*

Creativity and Economic Progress (New York: Oxford University Press, 1990), pp. 262, 263.

Page 50: On the flowering of research science as a tool of American industry, see Leonard Reich, *The Making of American Industrial Research: Science and Business at GE and Bell, 1876–1926* (Cambridge: Cambridge University Press, 1985), pp. 11, 12. On the cultural role of the engineer in the early twentieth century, see Cecilia Tichi's seminal work *Shifting Gears: Technology, Literature, Culture in Modernist America* (Chapel Hill, N.C.: University of North Carolina Press, 1987), particularly p. 98. For H. L. Mencken's quote on the engineering phenomenon, see Ibid., pp. 116–117.

Page 51, par. 1: On Charles Steinmetz's emblematic role see John M. Jordan, "Society Improved the Way You Can Improve a Dynamo: Charles Steinmetz and the Politics of Efficiency," *Technology and Culture,* January 1989.

Pages 51–53 (and next chapter): on the suburbs, regarding the English origins of the suburb, see Robert Fishman, *Bourgeois Utopias: The Rise and Fall of Suburbia* (New York: Basic Books, 1987), Ch. 1: The account of suburban life relies particularly on several excellent recent books and articles, including (in addition to *Bourgeois Utopias*) John R. Stilgoe, *Borderland: Origins of the American Suburb, 1820–1939* (New Haven: Yale University Press, 1988); and Mary Corbin Sies, "The City Transformed: Nature, Technology and the Suburban Ideal, 1877–1917," *The Journal of Urban History,* November 1987, pp. 81–111. Quote on page 00, Ibid., p. 83.

Page 52, par. 2: Lewis Mumford's comparison of city and suburb is drawn from his *The City in History: Its Origins, Its Prospects, Its Transformations and Its Prospects* (New York: Harcourt Brace Jovanovich, 1961), pp. 485–486.

Page 52, par. 3: The discussion of technology and suburban life draws particularly on Sies, "The City," pp. 89, 92, 103.

Page 53–54 (and next chapter): The discussion of the auto draws on a rich literature, including David L. Lewis and Lawrence Goldstein, eds., *The Automobile and American Culture* (Ann Arbor: University of Michigan Press, 1983); James J. Flink, *The Automobile Age* (Cambridge, Mass.: MIT Press, 1988); and Martin V. Melosi, *Coping with Abundance: Energy and Environment in Industrial America* (New York: Alfred A. Knopf, 1985).

Page 54, par. 2: The Muncie newspaper is quoted in Robert S. Lynd and Hellen Merell Lynd, *Middletown in Transition: A Study in American Culture* (New York: Harcourt Brace & World, 1956), p. 88.

Page 54, par. 3: The critic of consumer credit is quoted in Martha L.

Olney, "The Rise of Consumer Credit," a paper prepared for the Economic History Seminar, University of California—Berkeley, November 25, 1986, p. 17.

Page 55, par. 1: Tocqueville, *Democracy,* vol. 2, p. 239.

Page 55, par. 2: Calvin Coolidge quote in Joseph J. Seldin, *The Golden Fleece: Selling the Good Life to Americans* (New York: The Macmillan Co., 1963), p. 23.

Page 56 par. 1: Vance Packard, *The Waste Makers* (New York: David McKay, 1960), p. 29. Here and in the next chapter, the discussion of Americans' turn toward "self-fulfillment" draws heavily on the pathbreaking work of Richard Wightman Fox and T. J. Jackson Lears, *The Culture of Consumption: Critical Essays in American History, 1880–1980* (New York: Pantheon Books, 1983).

Page 56, par. 4: Ibid., p. 3. The Muncie newspaper is quoted in Lynd and Lynd, *Middletown,* p. 88.

Page 56, par. 5: On advertising and American culture, see Seldin, *Golden Fleece.* In par. 0: on advertising as a suggested response to Bolshevism, see Stewart Ewen, *All Consuming Images* (New York: Basic Books, 1988), p. 88.

Pages 58–59: The relationship between the garbage crisis of the late nineteenth century and the Progressive movement is brilliantly depicted in *Garbage in the Cities* (see p. 294).

Page 59, par. 2: On Ely's epiphany, see James Allen Smith, *Think Tanks and the Rise of the New Policy Elite* (New York: Free Press, 1990), p. 29. For an excellent account of swings in American ideology (discussed here and in the next chapter), see John B. Judis, "The Old and New Orders," *The American Enterprise,* July–August 1992, pp. 38–49.

Page 60, par. 3: Ronald Reagan, *An American Life: The Autobiography* (New York: Simon & Schuster, 1990), p. 42.

Chapter 3: America the Bountiful

Page 65, par. 2: Tocqueville, *Democracy,* vol. 2, p. 42.

Page 65, par. 4: On Thomas Jefferson's gadgets, see Jack McLaughlin, *Jefferson and Monticello* (New York: Henry Holt and Company, 1988), pp. 370–374. Leo Marx remarks on Jefferson's arcadian view of the machine in his seminal study of technology and culture, *The Machine in the Garden: Technology and the Pastoral Ideal in America* (New York: Oxford University Press, 1972), pp. 146–150 and 236.

Pages 66–68: The account here of eighteenth-century material culture draws heavily on *More Work,* Ch. 2, particularly pp. 17–27, 30–33. Another excellent source, though more useful for later periods, is Glenna Matthews, *Just a Housewife: The Rise and Fall of Domesticity in America* (New York: Oxford University Press, 1987). Detail on household implements is also drawn from Rolla Milton Tryon, *Household Manufacturers in the United States, 1640–1860,* quoted in Earl Lifshey, *The Housewares Story: A History of the American Housewares Industry* (Chicago: National Housewares Manufacturer's Association, 1973), p. 77–82.

Page 67: Recipe is from *More Work,* p. 22.

Page 67, par. 3, 4: On cleaning techniques, see Ibid., pp. 23–25, and *Just a Housewife,* p. 5, 6.

Page 67, par. 5: The account of early energy and energy technologies draws particularly on the excellent study by Louis C. Hunter, *A History of Industrial Power in the United States, 1780–1930,* vol. 2, *Steam Power* (Charlottesville, Va.: University Press of Virginia, 1985), and Martin Melosi, *Coping with Abundance: Energy and Environment in Industrial America* (New York: Alfred A. Knopf, 1985).

Page 68, par. 2: Kips Bay in the mid–nineteenth century is pictured in David Freeman Hawke, *Everyday Life in Early America* (New York: Harper & Row, 1988). See p. 242 for Gene Schermerhorn's account of his boyhood.

Page 68, par. 4: Washington story from Harold L. Burstyn, "What Can the History of Technology Contribute to Our Understanding?" in George Bugliarello and Dean B. Doner, eds., *The History and Philosophy of Technology* (Urbana, Ill.: University of Illinois Press, 1979), p. 57.

Page 69, par. 2: Harriet Martineau, quoted in Daniel J. Boorstin, *The Americans: The National Experience* (London: Weidenfeld & Nicolson, 1966), p. 25.

Page 69, Par. 3: On the transition to steam, see Hunter, *Industrial Power,* p. 109, and Melosi, *Coping with Abundance,* p. 23. On the technologies that came with steam, see Hunter, particularly pp. xxi, 252.

Page 69, par. 3: J. Blunt, "The Coal Business of the United States," *Hunt's Merchant's Magazine* 4 (1841), and *Scientific American,* May 21, 1853, both quoted in Hunter, *Industrial Power,* pp. 410–411.

Par. 71, par. 2: On the demographic transitions affecting American cities in the nineteenth century and the victory of New York over Calcutta, see *Garbage in the Cities* (see page 294), pp. 17–18.

Page 72, par. 2: On the consequences for everyday life of the Industrial Revolution, see *More Work,* pp. 40–41.

Page 72, par. 4: Charles Dickens, quoted in Thomas C. Cochran, *Challenge to American Values: Society, Business and Religion* (New York: Oxford University Press, 1985), p. 45. On figures for infant mortality, see Lewis Mumford, *The City in History: Its Origins, Its Transformations and Its Prospects* (New York: Harcourt Brace Jovanovich, 1961), p. 467.

Page 73, par. 1: The description of nineteenth-century Pittsburgh is from Hawke, *Everyday Life,* p. 243.

Page 73, par. 2: On the figures for production and output, see Ibid., p. 414. On the census of 1880, see Edmund C. Kirkland, *Industry Comes of Age: Business, Labor and Public Policy, 1860–1892,* vol. 4, *The Economic History of the United States* (New York: Holt, Rinehart & Winston, 1961), p. 171.

Page 73, par. 3: Mumford, *City in History,* p. 459.

Page 74, par. 3: On the changing shape of American industry, see Louis Galambos and Joseph Pratt, *The Rise of the Corporate Commonwealth, U.S. Business and Public Policy in the Twentieth Century* (New York: Basic Books, 1988), and Alfred D. Chandler, Jr., *Strategy and Structure: Chapters in the History of the Industrial Enterprise* (Cambridge, Mass.: MIT Press, 1962).

Page 74, par. 3: On nineteenth-century spending patterns, see Daniel Horowitz, *The Morality of Spending: Attitudes Toward the Consumer Society in America* (Nashville: University of Tennessee Press, 1987), particularly pp. 14, 21.

Pages 74 and 75, footnote, p. 75: On middle-class life, see Mathews, *Just a Housewife,* and *More Work.*

Pages 75–77: On the evolution of urban infrastructure, see Joel A. Tarr and Gabriel Dupuy, *Technology and the Rise of the Networked City in Europe and America* (Philadelphia: Temple University Press, 1988), particularly pp. 159–163. See also Hunter, *Industrial Power,* pp. 510–515. On appliances, see Melosi, *Coping with Abundance,* pp. 56, 59–60.

Page 77, par. 1: Tocqueville, *Democracy,* vol. 2, p. 43.

Page 77, par. 2: The account of the modernization of the American hotel is from Boorstin, *The Americans,* pp. 134–138.

Page 77, par. 3ff: On the evolution of the modern apartment house, see *Just a Housewife,* pp. 16–17.

Page 78, par. 2: On details regarding the development of bathroom facilities, see Tarr and Dupuy, *Technology,* p. 162; and Siegfried Giedion, *Mechanization Takes Command: A Contribution to Anonymous History* (New York: W. W. Norton, 1948), p. 685.

Page 78, par. 3: Regarding buying power, see Horowitz, *Morality of Spending,* pp. 19, 20; *More Work,* p. 61; and Sheila M. Rothman, *Woman's Proper Place* (New York: Basic Books, 1978), p. 18. On the stock of the Philadelphia hardware store, see Matthews, *Just a Housewife,* p. 12.

Pages 79–80: The account of the history of American merchandising draws on Lifshey, *Housewares Story,* pp. 80–93, 118–121; also Robert Hendrickson, *The Grand Emporiums* (New York: Stein and Day, 1979), pp. 34–36.

Pages 80–81: This account of the department store draws heavily on Matthews, *Just a Housewife,* pp. 17–23.

Pages 81–82: This account of suburban life draws heavily on Matthews, Cowan, *Just a Housewife;* Robert Fishman, *Bourgeois Utopias: The*

Rise and Fall of Suburbia (New York: Basic Books, 1987); John R. Stilgoe, *Borderland: Origins of the American Suburb, 1820–1939* (New Haven: Yale University Press, 1988); and Mary Corbin Sies, "The City Transformed: Nature, Technology and the Suburban Ideal, 1877–1917," *The Journal of Urban History,* November 1987, pp. 81–111.

Pages 82–84: The discussion of the auto draws on a literature that includes David L. Lewis and Lawrence Goldstein, eds., *The Automobile and American Culture* (Ann Arbor: University of Michigan Press, 1983); James J. Flink, *The Automobile Age* (Cambridge, Mass.: MIT Press, 1988); and Martin V. Melosi, *Coping with Abundance: Energy and Environment in Industrial America* (New York, Afred A. Knopf, 1985). On *Page 82, par. 3:* On Ford's mass production process, see Mark S. Foster, "The Automobile and the City," in Lewis and Goldstein, *The Automobile,* pp. 28, 29; and Melosi, *Coping with Abundance,* p. 106.

Page 83, par. 1: Sinclair Lewis, *Babbitt* (New York: Grosset & Dunlap, 1922), p. 24.

Pages 83–84: pars. 0–0: On the central role of the auto in the American economy, see Melosi, *Coping with Abundance,* pp. 106–109; and Reynold M. Wik, "The Early Automobile and the American Farmer," in Lewis and Goldstein, *The Automobile,* p. 43.

Page 84, par. 1: On the universal American embrace of the auto, see Martha L. Olney, "The Rise of Consumer Credit," a paper prepared for the Economic History Seminar, University of California—Berkeley, November 25, 1986, p. 11; Robert S. Lynd and Helen Merell Lynd, *Middletown in Transition: A Study in American Culture* (New York: Harcourt Brace & World, 1956), pp. 10, 265–266.

Page 85, par. 1: The discussion of wages, hours, and prices draws on Melosi, *Coping with Abundance,* pp. 100–103; Stanley Lebergott, *The American Economy: Income, Wealth and Want* (Princeton: Princeton University Press, 1975), p. 163; Horowitz, *Morality of Spending,* particularly pp. xxvi, 110; and *More Work,* p. 93.

Page 86, par. 1, 2: On consumer credit, see Olney, "Consumer Credit," particularly pp. 2–4.

Page 86, par. 4: Ellsworth quoted in *The Golden Fleece* (see page 294), p. 56. Giedion, *Mechanization,* p. 40.

Page 87, par. 1: Sinclair Lewis, *Main Street* (New York: Harcourt Brace & Company, 1920), p. 267.

Page 87–89: On advertising, see Joseph Seldin, *The Golden Fleece: Selling the Good Life to Americans* (New York: The McMillan Co., 1963), particularly pp. 194–195, 257; and Robert J. Samuelson, "The End of Advertising," *WP,* Aug. 14, 1991.

Page 89, par. 2, regarding the eradication of the Los Angeles mass transit system, see Fishman, *Bourgeois Utopias,* p. 156.

Page 89–90: The discussion of housewifely expertise draws on *More Work;* Rothman, *Woman's Proper Place;* Marilyn Ferris Motz and Pat Browne, eds., *Making the American Home: Middle Class Women and Domestic Material Culture 1840–1940* (Bowling Green, Ohio: Bowling Green State University Press, 1988); Susan Strasser, *Never Done* (New York: Pantheon Books, 1982); and *Just a Housewife,* particularly pp. 145–146 and 188–189.

Page 90: Christine Frederick, *Selling Mrs. Consumer* (New York: The Business Bourse, 1929), pp. 4–5, 13, 44, 52, 80–83, 201, 245, 251.

Chapter 4: Wealth and Waste

Pages 92–95: The discussion of U.S. GNP, per capita incomes, and consumption patterns during the postwar period draws on David M. Potter, *People of Plenty: Economic Abundance and the American Character* (Chicago: University of Chicago Press, 1954), particularly p. 83; Karl W. Deutsch, *Nationalism and Social Communication* (New York: John Wiley & Sons, 1953), pp. 36–45; Stanley Lebergott, *The American Economy: Income, Wealth and Want* (Princeton: Princeton University Press, 1975), p. 164; Samuel Bowles and Herbert Gintis, *Schooling in Capitalist America: Educational Reform and the Contradictions of Economic Life* (New York: Basic Books, 1977) p. 201, Joseph J. Seldin, *The Golden Fleece: Selling the Good Life to Americans* (New York: The Macmillan Co., 1963), pp. 43–47; and *More Work* (see page 294) pp. 192–196. The 1990 census data is reported in *NYT,* May 27, 1992, p. 17.

Page 96, footnote: See "Harper's Index," *Harper's Magazine,* January 1992, p. 13.

Page 97, par. 4: Kessinger, quoted in Michael Marriott, "In High Tech Dorms a Call for Power," *NYT,* April 12, 1991.

On page 98: Nixon, quoted in an unpublished paper by Sabrina Burghi, "The Paradox of Working Women in the World War II Decade," Columbia University, April 1992, p. 16.

Pages 98–99: The discussion of women's domestic vocation draws heavily on Betty Friedan, *The Feminine Mystique* (New York: W. W. Norton, 1963), pp. 60–65, 225.

Page 98, par. 4: Stevenson, quoted in Ibid., p. 60.

Page 99, par. 1: For Friedan on "things," see Ibid., p. 65. For the market research report, see Ibid., p. 225.

Page 99, par. 3: The discussion of working mothers draws mainly on the "America" survey, *The Economist,* Oct. 26, 1991, p. 6.

Page 100, par. 1: On white goods, see "Recycled Facts," *Garbage,* November–December 1991, p. 14.

Page 100, par. 3: Regarding the distribution of wealth in the early 1990s, see Sylvia Nasar, "The Rich Get Richer, But Never the Same

Way Twice," *NYT,* Aug. 16, 1992. On savings patterns, see "The Simple Life," *Time,* April 8, 1991, p. 63. Figures on the sneakers market come from Glenn Rifkin, "High Tops: High Style, High Tech, High Cost," *NYT,* Jan. 5, 19; and Patricia Leigh Brown, "The Once-Lowly Sneaker Is Pedestrian No More," *NYT,* May 28, 1992.

Page 102, par. 2: The discussion of the petrochemical industry and of plastics draws heavily on Peter H. Spitz, *Petrochemicals: The Rise of an Industry* (New York: John Wiley & Sons, 1988), particularly pp. 141–144, 120–121, 227–229, 231, 311. See also *War on Waste* (see page 000), pp. 13, 230.

Page 103, par. 5: Regarding makeup of the packaging waste stream, see The Plastics Recycling Foundation and the Center for Plastics Recycling Research, *Plastics Recycling: From Vision to Reality* (New Brunswick: Rutgers State University of New Jersey, 1988). On food waste, see EPA, *Characterization of Municipal Solid Waste in the United States: 1990 Update, Executive Summary,* June 1990, pp. ES-3–ES-6.

Page 104, par. 4: The data on farm chemicals is drawn from Susan Gilbert, "America Tackles the Pesticide Crisis," *The Good Health Magazine, The New York Times Magazine* (Part Two), Oct. 8, 1989; EPA, Pesticides Industry Sales and Usage, Fall 1992, pp. 24–27.

On *page 105, par. 4:* On the Great Lakes as water source, see Pendleton Scott, "Toxics #1 Challenge for Great Lakes Basin," *The Christian Science Monitor,* Jan. 9, 1991; also GAO, *Pesticides in the Great Lakes,* Washington, D.C., June 1993.

Page 106, par. 2: On concentrations of PCBs, see EPA, *Pollution Prevention 1991: Progress on Reducing Industrial Pollutants,* October 1991, p. 195.

Page 107, par. 1: On the relationship between growth of population and garbage output after World War I, see *Garbage in the Cities* (see page 294), pp. 1, 192, where Melosi cites *Cities and the Nation's Disposal Crisis,* National League of Cities and United States Conference of Mayors, Solid Waste Management Task Force; and Anne Erlich and Paul Erlich, *Healing the Planet,* (Reading, Mass.: Addison-Wesley, 1991), p. 142. On the comparison of U.S., German and Japanese garbage totals, see Bette K. Fishbein, *Germany's Packaging Law: A Nation Confronts Its Solid Waste Crisis.* (New York: INFORM, 1994) pp. 6–7. On estimates of wastes released into U.S. coastal waters and growing sludge problems, see "Sewage Pipe Break Closes San Diego Beaches," *Biocycle,* March 1992, p. 55, and GAO, *Water Pollution: Serious Problems Confront Emerging Municipal Sludge Management Program,* March 1990, pp. 8, 15, 32–34.

Page 107, par. 4: On mining wastes, see John E. Young, *Mining the Earth,* WorldWatch Paper 109 (Washington, D.C.: WorldWatch Institute, July 1992), pp. 16–27. On pollution from mining and forestry, see GAO, *Water Pollution*, March 1990, p. 9.

Page 108, par. 3: On the industrial waste stream, see GAO, *Nonhazardous Waste: Environmental Safeguards for Industrial Facilities Need to Be Developed*, April 1990, p. 18; and OTA, *Hazardous Waste: Funding of Post-closure Liabilities Remains Uncertain*, June 1990, p. 8; Jim Glenn, "Recycling at the Factory Gate," *Biocycle*, June 1993, p. 42.

Page 108, footnote: see John E. Young, *WW*, March–April 1992, pp. 26–33.

Page 109: On production of synthetic chemicals and petrochemicals, see Mark H. Dorfman, Warren R. Muir, and Catherine G. Miller, *Environmental Dividends: Cutting More Chemical Waste* (New York: INFORM, 1992); Amy Martin, "A Petrochemical Primer," *Garbage*, September–October 1991, p. 38; and April 1990, p. 19.

Pages 109–110: On the use of chemicals in agriculture, see Peter Weber, "A Place for Pesticides?" *WW*, May–June 1992; Peter H. Spitz, *Petrochemicals: The Rise of an Industry* (New York: John Wiley & Sons, 1988), p. 311; Lester R. Brown and John E. Young, "Feeding the World in the Nineties," in *The State of the World 1990* (Washington, D.C.: WorldWatch Institute, 1991); GAO, *Alternative Agriculture: Federal Incentives and Farmers' Options*, February 1990; Benjamin A. Goldman, *The Truth About Where You Live* (New York: Times Books, 1991), p. 223; Evan Eisenberg, "Back to Eden," *The Atlantic*, November 1989, p. 59; "Harper's Index," *Harper's Magazine*, August 1989, p. 15; and Lester Brown, "Fertilizer Engine Losing Steam," *WW*, September–October 1991, pp. 32–33; GAO, Lawn Care, *Pesticides Reregistration May Not Be Completed Until 2006*, Washington, D.C., May 1993; *FAO Fertilizer Yearbook*, United Nations Food and Agricultural Organization (Rome: The United Nations, 1992).

Page 111, par. 3: Irving Shapiro, quoted in "End of Partnership."

Pages 111–114: On nuclear wastes, see Scott Saleska, "Low-level Radioactive Waste: Gamma Rays in the Garbage," *BAS*, April 1990, p. 20; Nicholas Lenssen, *Nuclear Waste: The Problem That Won't Go Away*, WorldWatch Paper 106 (New York: WorldWatch Institute, December 1991); Nicholas Lenssen, "Facing Up to Nuclear Waste," *WW*, March–April 1992, p. 11; Matthew L. Wald, "End of Partnership Forged in Birth of Atomic Age," *NYT*, Oct. 16, 1988; "The Contamination Factory," *BAS*, October 1991, p. 34; and COTA, *Complex Cleanup: The Environmental Legacy of Nuclear Weapons Production*, February 1991, p. 2.

Page 111, par. 2: On nonnuclear hazardous waste, see particularly GAO, *Hazardous Waste: Improvements Needed in DOD's Contracting System for Disposal of Waste*, August 1991; and Keith Schneider, "Military Has New Strategic Goal in Cleanup of Vast Toxic Waste," *NYT*, Aug. 5, 1991.

Page 113 par. 2: Regarding the technicians' joke about spent fuel, see Bill Breen, "Dismantling Nuclear Power Plants," *Garbage,* March–April 1992, p. 47.

Page 114, par. 2: The account of decommissioning a nuclear plant leans heavily on Cynthia Pollock Shea, "Breaking Up Is Hard to Do," *WW,* July–August 1989, pp. 10–11. See also Shea, *Decommissioning: Nuclear Power's Missing Link,* WorldWatch Paper 69 (New York: WorldWatch Institute, 1986); and Breen, "Dismantling," p. 47.

Chapter 5: Swamped

Pages 116–117: On the definition and extent of nonpoint pollution, see GAO, *Water Pollution: Greater EPA Leadership Needed to Reduce Nonpoint Source Pollution,* October 1990; Oliver A. Houck, "America's Mad Dash to the Sea," *The Amicus Journal,* Summer 1988; GAO *Alternative Agriculture: Federal Incentives and Farmers' Options,* February 1990, p. 15; GAO, *Pesticides: EPA Could Do More to Minimize Groundwater Contamination,* April 1991, pp. 14–15; and Benjamin A. Goldman, *The Truth About Where You Live* (New York: Times Books, 1991), pp. 223, 232–233, where he cites Elizabeth B. Nielsen and Linda K. Lee, *The Magnitude and Costs of Groundwater Contamination from Agricultural Chemicals: A National Perspective* (Washington, D.C.: National Technical Information Service, 1987).

Pages 118–120: On nuclear wastes, see Scott Saleska and Arun Makhijani, "Hanford Cleanup: Explosive Solution," *BAS,* October 1990, pp. 14–20; Matthew L. Wald, "38-Year Plutonium Loss at Plant Equals 7 Bombs," *NYT,* March, 3, 1990; Bryan Abas, "Rocky Flats: A Big Mistake from Day One," *BAS,* December 1989, pp. 19–24; Cass Peterson, "Energy Department Cleanup Could Cost $128 Billion," *WP,* Jan. 5, 1989; Keith Schneider, "Public Plan Asked on Atomic Cleanup," *NYT,* December 15, 1988.

Page 120, par. 1: On cleanup costs, see Jessica Tuchmann Mathews, "Superfund Boondoggle," *WP,* Sept. 6, 1991; Robert Alvarez and Arun Makhijani, "Nuclear Waste: the $100 Billion Mess," *WP,* Sept. 4, 1988; and J. W. Maurits la Rivière, "Threats to the World's Water," *Scientific American,* September 1989, pp. 29, 30.

Page 120–121: This account of pollution from landfills draws most directly on the seminal analysis of waste disposal options in Richard A. Denison and John Ruston, *Recycling and Incineration* (Washington, D.C.: Island Press, 1990), p. 5; GAO, *State Management of Municipal Landfills and Landfill Expansions,* June 1989; and Bill Breen, "Landfills are #1," *Garbage,* September–October 1990, pp. 42–47.

Page 121–122: On pollution from incineration, see Denison and Ruston, *Recycling and Incineration,* pp. 63–65, and 177–97; Jerome O. Nriagu, "Industrial Activity and Metal Emissions," a paper presented at the Global Change Institute Conference on Industrial Ecology and Global Change, Snowmass, Colorado, July 20–31, 1992, p. 4 [in R. Socolow et al., eds., *Industrial Ecology and Global Change* (New York: Cambridge University Press, 1994)]; also "Fire and Ice: How Garbage Incineration Contributes to Global Warming," a report prepared by Elizabeth Holtzman, Office of the Controller, City of New York, March 1992; and Hunter F. Taylor, P.E., "Municipal Waste-to-Energy Facilities Reduce Greenhouse Gas Emissions," a paper presented at the Institute of Gas Technology Fourth Annual National Symposium on Municipal Solid Waste Disposal and Energy Production, Jan. 16, 1990, pp. 7–9.

Page 122–123: On oil wastes from shipping, see John Grassy, "The Waste Oil Monster," *Garbage,* July/August 1991, p. 34; Maurits la Rivière, "Threats," p. 89; "Vital Signs," and "Vital Signs" *WW,* February 1992, p. 6. On recapitulating effects, see Jon R. Luoma, "Some Oil Spills Repeat Harm Again and Again," *NYT,* Dec. 21, 1993.

Pages 122–124: On nonpoint oil pollution, Eric Schmitt, "Gas Spill Cleanup Turns L.I. Backyards into Field Labs," *NYT,* May 14, 1990; and Grassy, "Waste Oil Monster," p. 36.

Page 124: The story of oil pollution in Arthur Kill is from Katherine C. Parsons, "The Birds: A Harbor-Watcher's Diary," *Garbage,* November/December, 1991, pp. 38–43.

Pages 122–128: On industrial waste, see GAO, *Nonhazardous Waste: Environmental Safeguards for Industrial Facilities Need to Be Developed,* April 1990; OTA, "Managing Industrial Solid Wastes from Manufacturing, Mining, Oil and Gas Production and Utility Coal Combustion," Washington, D.C., March, 1992. Houck, "America's Mad Dash"; GAO, *Hazardous Waste: Funding of Postclosure Liabilities Remains,* June 1990; "Toxic Data Flow," *Environmental Action,* September–October 1989; GAO, *Toxic Substances: Status of EPA's Reviews of Chemicals Under the Chemical Testing Program,* October 1991; GAO, *Hazardous Waste: Limited Progress in Closing and Cleaning up Contaminated Facilities,* May 1991.

Pages 128–129: On Superfund and cleanup, see GAO, *Hazardous Waste: Status and Resources of EPA's Corrective Action Program,* April 1990, p. 17; *Superfund: More Settlement Authority and EPA Controls Could Increase Cost Recovery,* July 1991, p. 9; and GAO, *Superfund: Progress, Problems, and Reauthorization Issues,* Statement of Richard Hemtora, April 21, 1993.

Pages 130–131: On surface waters, see Maurits la Rivière, "Threats"; Roberta Savage, "Water Continues to Improve Under the Clean Air Act," *Chemecology,* October 1991; and Michael Specter, "The World's Oceans Are Sending an S.O.S.," *NYT,* May 3, 1992.

Pages 131–132: On groundwater pollution, see *Environmental Almanac,* compiled by World Resources Institute (Washington, D.C., 1992), pp. 90–91, citing the Isaak Walton League study of groundwater protection; William H. MacLeish, "Water, Water Everywhere, How Many Drops to Drink?" *World Monitor*, December 1990; Maurits la Rivière, "Threats," pp. 80, 86; Elizabeth Kolbert, "Ugly Reminders of the Oil That Went Underground," *NYT*, August 22, 1990; and Mary E. Tiemann, "Groundwater Contamination and Protection," CRS Issue Brief, Dec. 27, 1988.

Page 133: Regarding impacts on fish species of polluting runoffs, see B. Drummond Ayres, Jr., "Stirrings of Hope in Redeeming Chesapeake Bay," *NYT*, Dec. 2, 1990; Nicholas Lenssen, "Ocean Blues," *WW*, July–August 1989 pp. 26ff; William E. Schmidt, "In the Great Lakes, Some Pollution Defies the Cleanup," *NYT*, July 2, 1989; "From Minnow to Sturgeon, North American Fish Are in Peril," *NYT*, Jan. 30, 1990; Tyler Bridges, "Louisiana Environmentalists Set a Course for Saving Lake Pontchartrain," *NYT*, June 24, 1990.

Pages 134–135: On wetlands and coastal waters, see H. Jesse Walker, "The Coastal Zone," pp. 277ff; and S. C. Snedaker, "Mangroves: Their Value and Perpetuation," *Nature and Resources* 14, 3, pp. 6–13 in *Earth Transformed* (see p. 294).

Pages 135–137: On ocean pollution, see Timothy D. Jickells, Roy Carpenter, and Peter S. Liss, "Marine Environment," in *Earth Transformed*, p. 330; Lennsen, "Nuclear Waste," p. 30; John T. Hardy, "Where the Sea Meets the Sky," *Natural History*, May 1991, p. 59; and "Plenty of Fish in Sea? Not Anymore," *NYT*, March 25, 1992.

Chapter 6: Nowhere to Hide It

Pages 142–43: The account of the Fresh Kills landfill draws most directly on Bill Breen, "Landfills Are #1," *Garbage*, September–October 1990, p. 43.

Pages 143–144: On RCRA and the waste hierarchy, see *War on Waste* (see page 294), especially pp. 63–65; Brenda Platt et al., *Beyond 40 Percent* (Washington, D.C.: Island Press and the Institute for Local Self-Reliance, 1991), especially pp. 2–5; and J. Winston Porter, "Our Garbage Problem Won't Go Away By Itself," *Chemecology*, May 1989.

Page 145, par. 2: For figures on landfilling, incineration, and recycling, see "Characterization of Municipal Solid Waste in the United States 1992 Update," prepared for the U.S. Environmental Protection Agency by Franklin Associates, Ltd., July 1992; Robert Steuteville and Nora Goldstein, "The State of Garbage in America," *Biocycle*, May 1993, pp. 42–50; and Edward W. Repa and Susan K. Sheets, "Landfill Capacity in America," *Waste Age*, May 1993, pp. 19ff.

Page 145, par. 3: On figures for recycling, see Steuteville and Goldstein, "The State of Garbage," pp. 47–48.

Page 145, par. 3: For data on tipping fees, see "The 1987 Tip Fee Survey, National Solid Wastes Management Association, Technical Bulletin 88-1, Jan. 30, 1988; and Bette K. Fishbein and Caroline Gelb, *Making Less Garbage: A Planning Guide for Communities* (New York: IN-FORM, 1993), pp. 15–16. Philadelphia's extraordinary rise in rates is noted in *War on Waste,* p. 125.

Page 146, par. 3: On landfill closings and capacity, see Repa and Sheets, "Landfill Capacity," p. 19; *Rubbish!* (see esp. pp. 92–94, 106–109); and Jeff Bailey, "Economics of Trash Shift as Cities Learn Dumps Aren't So Full," *WSJ,* June 2, 1992.

Page 147, par. 8: On states' enforcement of landfill standards, see Richard A. Denison and John Ruston, *Recycling and Incineration* (Washington, D.C.: Island Press, 1990), especially pp. 4–6; and Robert Pear, "U.S. Sets Rules to Cut Landfill Pollution," *NYT,* Sept. 12, 1991.

Pages 149–151: The history of incineration in the United States draws principally on *Garbage in the Cities* (see page 294), pp. 170–172; *War on Waste,* pp. 7–8, 39–48, 165, 171–173; *Trash to Cash* (see page 294), pp. 51–52, 241–242; Platt et al., *Beyond 40 Percent,* pp. 3–4; and Fishbein and Gelb, *Making Less Garbage,* p. 14. See also Staff of *Newsday* (Long Island), *Rush to Burn: Solving America's Garbage Crisis* (Washington, D.C.: Island Press, 1989), for accounts of several communities' experiences with garbage politics and the incineration industry.

Page 151: The story of LANCER draws heavily on Brumberg and Gottlieb's detailed analysis in *War on Waste,* pp. 155–188, which provides the central focus for their book.

Page 152, par. 1: For more detail on local conflicts between recycling and incineration, see Platt et al., *Beyond 40 Percent,* especially p. 4.

Page 152, par. 3: Regarding incineration's record on trash reduction, see Ibid., p. 4; and John E. Young, "Burn Out," *WW,* July–August 1991, p. 8.

Pages 152–154: On pollution from incineration, see especially Denison and Ruston *Recycling and Incineration,* and Richard A. Denison, "Health Risks of Municipal Solid Waste Incineration: The Need for a Comprehensive View," a paper prepared for the Rene Dubos Center Forum Program, Only One Earth Forum on the City as a Human Environment, New York, May 16–18, 1989, pp. 10–11, 19–20. See also "Fire and Ice: How Garbage Incineration Contributes to Global Warming," a report prepared by Elizabeth Holtzman, Office of the Controller, City of New York, March 1992. On actual performance of incinerators, see Bette K. Fishbein, "State of the Art and Actual Practice: Preliminary Findings from a Study of 15 Waste to Energy Plants," the Seventh Annual Conference on Solid Waste Management and Materials Policy, New York, Jan. 29–31, 1991.

Page 154, par. 2: On toxicity of incinerator ash, see Denison and Ruston, *Recycling and Incineration,* pp. 16–18, citing the 1898 Environmental Defense Fund data base; and *War on Waste,* p. 111.

Page 154, par. 3: On the tug-of-war between recycling and incineration, see Hannah Holmes, "Cities Fight for Right to Recycle," *Garbage,* September–October 1990, pp. 00–00. The Babylon and Claremont examples are drawn from Platt et al., *Beyond 40 Percent,* p. 10.

Chapter 7: Choosing Conservation

Page 158, par. 1: For an illuminating account of source reduction modalities for consumers see Bette K. Fishbein and Caroline Gelb, *Making Less Garbage: A Planning Guide for Communities* (New York: INFORM, 1993).

Page 158, par. 3: On polling and attitudes, see, for example, "Americans Prefer Recyclable Containers by More Than 3 to 1: Comprehensive Curbside Collection Gains Strong Support," press release by the Glass Packaging Institute, Washington, D.C., Jan. 17, 1990.

Pages 159–161: The history of recycling in the United States draws particularly on *Garbage in the Cities,* pp. 51–78; William J. Rathje, "The History of Garbage," *Garbage,* September–October 1990, p. 40; *War on Waste* (see page 294), pp. 195–200; Daniel C. Walsh, "Will We See a Repeat of 1918?" *Waste Age,* December 1989, pp. 27, 31–32; "Scrap: America's Ready Resource," *Phoenix Quarterly,* 20 (no. 1–2, 1988); Gert Modell, "Still Valid: 50-Year Lesson in Recycling," *Waste Age,* July 1990, pp. 102–104; Eugene J. Wingerter, "Where the Waste Industry Is Going," *Waste Age,* April 1990, pp. 281–284; and *Trash to Cash* (see page 294), pp. 2–3.

Page 161, par. 1: On recycling at the U.S. Post Office, see Scott L. Bashore and James M. Phillips, "Postal Service 'Soars' into Recycling," *Biocycle,* February 1992, pp. 76–77.

Page 161–162: On Burlington's recycling experience, see Emily Fleschner with George Crombie and Tom Moreau, "A City Shifts into Full-Scale Recycling," in *Biocycle,* January 1992, pp. 38–42. Other sources are interviews with Burlington officials, August and November 1993.

Pages 162–163: On Seattle's recycling experience, see Jerome Richard, "Better Homes and Garbage: Seattle, Recycling Capital of the Nation," *The Amicus Journal,* Summer 1990, pp. 50–51. The account also draws on interviews with Seattle officials, August 1993. Seattle's plan is described in Platt et al., *Beyond 40 Percent* (Washington, D.C.: Island Press and the Institute for Local Self-Reliance, 1991), pp. 184–185.

Page 163, par. 2: John E. Pepper quoted in John Holusha, "An Industry Tries to Improve Its Record on Plastic," *NYT,* March 31, 1991.

Pages 163–164: The account of families' reactions to recycling in

1989 is derived from Dena Kleiman, "A Simple Domestic Chore Becomes a Cause," *The NYT*, July 26, 1989.

Pages 165–171: The overview of current and potential marketing possibilities of recycled materials leans very heavily on Susan Williams's exhaustive investigation of commercial possibilities in the postconsumer waste stream in *Trash to Cash.*

On scrap steel, see particularly pp. 51–54; also, Kurt Smallberg, "Steel Can Recycling: The Future Is Today" (cited in *Trash to Cash,* p. 49), a presentation at the "Building Confidence in Recycling" conference sponsored by the U.S. Conference of Mayors, Washington, D.C., March 29, 1990.

On paper, see *Trash to Cash,* pp. 70–83; also Jerry Powell, "How Are We Doing? The 1991 Report," *Resource Recycling,* April 1992, p. 78; Steve Apotheker, "Market Trend for Old Newspapers," *Resource Recycling,* July 1992, pp. 25–29, 92; and Fred Iannazzi and Richard Strauss, "Changing Markets for Recycled Paper," *Resource Recycling,* April 1992, p. 84.

On glass, see *Trash to Cash,* particularly pp. 31–45; and Powell, "How Are We Doing," p. 74. On plastics, see *Trash to Cash,* particularly Ch. 5. On yard waste, see *Trash to Cash,* particularly pp. 204–209. On one plastic can, see *Trash to Cash,* p. 18. On pollution from recycling, see *Trash to Cash,* particularly pp. 36–37, 45, 73–74, 120–121.

Pages 171–173: On source reduction, see Fishbein and Gelb, *Making Less Garbage,* p. 3.

Page 173, par. 3: Walter R. Stahel, "Re-use and Re-Cycling: Waste Prevention and Resource Savings in Utilization," a paper prepared for Woods Hole (Massachusetts) Conference on Industrial Ecology, July 1992.

Chapter 8: Can Technology Fix It?

Page 178, par. 2: David Lowenthal, "Awareness of Human Impacts: Changing Attitudes and Emphases," in *Earth Transformed by Human Action:* (see page 294), p. 131.

Page 180, par. 2: "Device Helps Bars Recycle," *Recycling Times,* Aug. 15, 1989; and Eben Shapiro, "Empty Cans Go In; Coins and Coupons Come Out," *NYT,* June 10, 1992.

Pages 181–183: The account of plastics recycling technologies draws particularly on Peter Dinger, "Automatic Sorting for Mixed Plastics," *Biocycle,* March 1992, pp. 80–82; Peter Dinger, "Automatic Microsorting for Mixed Plastics," *Biocycle,* April 1992, pp. 79–80; *Trash to Cash* (see page 000), pp. 116–120.

Page 182, par. 2: For data on plastics baling and bulk, see Kathleen White, "MRF of the Month: The Milford Recycling Facility," *Waste Age,* May 1992, p. 47.

Page 184–185: On the quest for biological processes, see "Lime Said to Clean PCB Sites Easily," *NYT,* March 13, 1991; Matthew L. Wald, "A Company Is Finding the Good in Pollution," *NYT,* Jan. 9, 1991; "Scientists Discover Way to Use Sunlight to Dissolve Oil Slicks," *Chemecology,* May 1992; John Holusha, "New Techniques to Turn an Oil Spill into a Collectible," *NYT,* April 21, 1991; and Michael G. Malloy, "High-Tech Waste Treatments Set to Take Off," *Waste Age,* March 1992, pp. 79–80.

Page 184, par. 1: George Bugliarello describes the rising potential of biotechnology, in "Technology and the Environment," in Daniel B. Botkin, *Changing the Global Environment* (New York: Academic Press, 1989), pp. 389–391.

Page 184, par. 4: On worm composting, see Susan Combs, "Food Composting Made Easy," *Waste Age,* October 1990, pp. 117–118; Marj Charlier, "The Real Worry Is What Happens When They Finish All the Garbage," *WSJ,* March 9, 1989. On William Brinton, see David Stipp, "At Café Brinton, Today's Special Is Chicken à la Sawdust," *WSJ,* July 31, 1991.

Pages 185–187: On aquatic-plant treatment, see Janet Marinelli, "After the Flush: The Next Generation," *Garbage,* January–February 1990, pp. 24–35; also Doug Stewart, "Nothing Goes to Waste in Arcata's Teeming Marshes," *Smithsonian,* April 1990, pp. 175–179; Robert Spencer, "Lower Cost Way to Septage Treatment," *Biocycle,* March 1992, pp. 64–68; William MacLiesh, "Water, Water, Everywhere, How Many Drops to Drink?" *World Monitor,* December 1990, pp. 54–58; and Guada Woodring, "A 'Growing' Interest in Wastewater Plants," *Waste Age,* June 1990, pp. 87–92.

Page 187, par. 2: On Wolverton's plant filter, see Janet Marinelli, "After the Flush," p. 31.

Pages 187–188: On biotechnologies for cleanup, see John Douglas, "Cleaning Up with Biotechnology," *EPRI Journal,* September 1988, pp. 15–21; Clarence G. Golueke and Luis F. Diaz, "Bioremediation for Hazardous Wastes," *Biocycle,* February 1990, pp. 54–55; and Jerald Schnoor, "Soil as a Vulnerable Environmental System," a paper presented at the Global Change Institute Conference on Industrial Ecology and Global Change, Snowmass, Colorado, July 20–31, 1992.

Page 188, par. 1: On Houston, Texas case, see "Garbology," *The Economist,* Feb. 13, 1988, p. 82. On bioremediation at hazardous waste sites, see OTA, *Bioremediation for Marine Oil Spills,* May 1991, p. 1.

Pages 189–190: On biotechnologies in agriculture, see Jeanne McDermott, "Some Heartland Farmers Just Say No to Chemicals," *Smithsonian,* April 1990, pp. 114–127; Peter W. Tonge, "The No-Pesticide Revolution," *World Monitor,* June 1989, pp. 56–60; William K. Stevens,

"New Microbial Agents May Help Kill Crop Pests," *NYT,* July 10, 1990; Rhonda L. Rundle, "Genetic-Engineered Pesticides May Reach the U.S. Market Soon," *WSJ,* June 24, 1991; Evan Eisenberg, "Back to Eden," *The Atlantic,* November 1989, pp. 57–89; and Malcolm Gladwell, "Warding Off Insect Pests with Bioengineering," *WP,* April 17, 1989.

Page 190, footnote: Regarding engineering new genetic material, see "The Virtues of Starvation," *The Economist,* April 28, 1990, p. 93; and William K. Stevens, "Tests Speed Effectiveness of Biological Pesticide," *NYT,* July 5, 1991.

Pages 191–192: Regarding biotechnologies for plastics manufacture, see William D. Luzier, "Materials Derived from Biomass/Biodegradable Materials," *PNAAS* 89 (February 1992), pp. 839–842; R. Clinton Miller and Robert W. Lenz, "Natural Plastics," *Natural History,* May 1990, pp. 82–84; John Holusha, "Scientists Are Proving That Natural Plastic Is Not an Oxymoron," *NYT,* Oct. 21, 1990; and "Plastic from a Plant," *NYT,* April 28, 1992.

Pages 192–193: On dematerialization, see Jesse H. Ausubel and Hedy E. Slavovich, eds., *Technology and Environment* (Washington, D.C.: National Academy of Engineering/National Academy Press, 1989), pp. 56ff.; Marc Ross, "Efficient Energy Use in Manufacturing," *PNAAS* 89 (February 1992), p. 827; Eric D. Larson, Marc H. Ross, and Robert H. Williams, "Beyond the Era of Materials," *Scientific American,* June 1986, pp. 34–41; OTA, *Green Products by Design: Choices for a Cleaner Environment,* October 1992; "Energy Enigmas," in "Lifting the Lid," *Garbage,* July–August 1992, p. 14; Glenn Rifkin, "Smart Plans for Clogged Roads," *NYT,* Nov. 20, 1992; Robert Herman, Simiak A. Ardekani, and Jesse H. Ausubel, "Dematerialization," in Ausubel and Slavovich, *Technology and Environment,* especially pp. 52–53; Walter R. Lynn, "Engineering Our Way Out of Endless Environmental Crises," in Ibid., pp. 184–185; and Holly Brough, "Jet Travel Soars," *Worldview,* May–June 1992, p. 36; Braden Allenby, "Industrial Ecology: The Materials Scientist in an Environmentally Constrained World" *MRS Bulletin,* March 1992, pp. 49–50.

Pages 194–195: On new materials, see Jonathan P. Hicks, "Light and Tough, New Materials Head for the Mass Marketing Stage," *NYT,* March 25, 1990; John Holusha, "Layer by Layer to the Perfect Blend of Metals," *NYT,* Dec. 1, 1991; "The Right Stuff," *The Economist,* Nov. 24, 1990, pp. 95–97.

Pages 195–199: Regarding industrial ecology, this account draws particularly on the pioneering work of Robert U. Ayres, especially "Industrial Metabolism," in Ausubel and Slavovich, *Technology and Environment* (Washington, D.C.: National Academy of Engineering, 1989), pp. 23–49; and *PNAAS* 89 (February 1992), pp 815–820. See also papers from the Global Change Institute Conference on Industrial Ecology and Global Change, Snowmass, Colorado, July 20–31, 1992; and papers from the Woods Hole (Massachusetts) Conference on Industrial Ecology, July 1992. See also Hardin B. C. Tibbs, "Industrial Ecology—An Agenda for Environmental Management," *PPR*, Spring 1992, especially pp. 171–172 on the Kalendborg waste exchange; and Robert A. Frosch and Nicholas E. Gallopoulos, "Strategies for Manufacturing," *Scientific American*, September 1989, p. 149.

Page 198, par. 2: On the competitive advantages in green design, see Braden R. Allenby, "Design for Environment: A Tool Whose Time Has Come," *SSA Journal*, September 1991, pp. 5–9.

Page 198, par. 3: On industrial source reduction cases, see Karl DeWahl and Donna Peterson, "Waste Reduction in Solvent Cleaning: Process Changes Versus Recycling," *PPR*, Winter 1991–92, p. 74; EPA, *Pollution Prevention 1991: Progress on Reducing Industrial Pollution*, Washington D.C., 1991, p. 51; Jeffrey R. Adrian, "Managing Solvents and Wipers," *PPR*, Autumn 1991, pp. 419–423; Elizabeth H. Rose and Arthur D. Fitzgerald, "Free in Three: How Northern Telecom Eliminated CFC-113 Solvents from Its Global Operations," *PPR*, Summer 1992, pp. 302–306; and F. William Kirsch and Gwen P. Looby, "Pollution Prevention in Practice," *PPR*, Spring 1991, pp. 162–164.

Page 198, footnote: See Kenneth E. Nelson, "Practical Techniques for Saving Energy and Reducing Waste," paper prepared for 1992 Global Change Institute: Industrial Ecology and Global Change, Snowmass, Colorado, July 19–31.

Page 200, par. 1: On packaging, see Larry J. Nielsen, "How Digital Uses Measurement Techniques in Managing Packaging Waste," *PPR*, Summer 1991, p. 250.

Pages 200–201: For an authoritative account of the German recycling system and the German "take-back" regulations see Bette K. Fishbein, *Germany's Packaging Law: A Nation Confronts Its Solid Waste Crisis* (New York: INFORM, 1994); also, Cynthia Pollock Shea, "Package Recycling Laws," *Biocycle*, June 1992, pp. 56–58.

Page 201, par. 4: On German "take-back" regulations, see also OTA, *Green Products by Design* (Washington, D.C., 1992), pp. 70–71.

Pages 201–202: On design for disassembly, I am indebted here to the analysis in Ibid., particularly pp. 39, 59; see also Doron P. Levin, "Imperatives of Recycling Are Gaining on Detroit," *NYT*, Sept. 12, 1992.

Chapter 9: Can Science Sort It Out?

Page 204, par. 3: The EPA quote on costs and benefits is from EPA, *Environmental Investments: The Cost of a Clean Environment,* December 1990, p. vii.

Pages 205–208: For the discussion of life-cycle analysis, I am particularly indebted to the Working Paper of Group III of the Global Change Institute Conference on Industrial Ecology and Global Change, Snowmass, Colorado, July 20–31, 1992, pp. 17–28. See also Barry Meier, "Life-cycle Studies: Imperfect Science," *NYT,* Sept. 22, 1990.

Pages 206–208: On life-cycle controversies, see Scott Chaplin, "The Return of Refillable Bottles," *Resource Recycling,* March 1991, pp. 132–133; Robert W. Fenton, "Reuse versus Recycling: A Look at Grocery Bags," *Resource Recycling,* March 1992, pp. 105–107; and Martin B. Hocking, "Paper versus Polystyrene: A Complex Choice," *Science,* February 1991, pp. 504–505. Regarding the conflict over cloth and disposable diapers, see particularly Patricia Poore, "Disposable Diapers Are OK," *Garbage,* October–November 1992, pp. 26–31.

Pages 208–215: Regarding risk assessment, for a comprehensive discussion of the issues see Richard N. L. Andrews, "Risk Assessment: Regulation and Beyond," in *Environmental Policy in the 1990s,* Norman J. Vig and Michael E. Kraft, eds. (Washington, D.C.: Congressional Quarterly Press, 1990), pp. 167–186.

Page 209: D. Alice Ottoboni, quote from her "The Dose Makes the Poison," *Garbage,* October–November 1992, pp. 38–43. The EPA scientists are quoted from GAO, *Water Pollution: Serious Problems Confront Emerging Municipal Sludge Management Program,* March 1990, pp. 42–43.

Pages 209–210: Regarding the Delaney clause, see GAO, *Toxic Substances: Effectiveness of Unreasonable Risk Standards Unclear,* July 4, 1990, p. 4; and Keith Schneider, "A Trace of Pesticide, an Accepted Risk," *NYT,* Feb. 7, 1993. On public perceptions of risk, see Lester B. Lave, "Risky Business: Thinking About the Benefits and Costs of Government Regulation," *The American Enterprise,* November–December 1992, p. 21; and *Public Attitudes Toward Garbage Disposal* (Washington, D.C.: National Solid Waste Management Association, 1988).

Pages 210–211: Regarding relative environmental risks, see Michael W. Pariza, "A New Approach to Evaluating Carcinogenic Risk," *PNAAS* 89 (Febuary 1992), pp. 860–861; EPA, *Reducing Risk: Setting Priorities and Strategies for Environmental Protection,* September 1990.

Pages 211, par. 1: On the comparison between expenditures on breast cancer and toxic cleanups, see Jocelyn White, "Superfund: Pouring Money Down a Hole," *NYT,* April 17, 1992.

Page 211, footnote: See Working Paper of Group II, 1992 Global Change Institute: Industrial Ecology and Global Change, Snowmass, Colorado, July 19–31.

Pages 211–212: Regarding uncertainties about the extent of risks, see GAO, *Toxic Substances: EPA's Chemical Testing Program Has Not Resolved Safety Concerns,* April 1991, p. 22; also the Working Paper of Group II of the Global Change Institute Conference on Industrial Ecology and Global Change, Snowmass, Colorado, July 20–31, 1992; Ann Misch, "Chemical Reaction, *WW*, March–April 1993, p. 13.

Page 212, par 2: EPA, quote from EPA, *Reducing Risk*, p. 13. On animal studies, see OTA, *Green Products*, pp. 91, 103; and Misch, "Chemical Reaction," p. 35.

Pages 212–213: Regarding the controversy over dioxin, see Ed Ayres, "Pursuing the Truth About Paper," *WW*, September–October 1992, pp. 17–25; also Keith Schneider, "Panel of Scientists Finds Dioxin Does Not Pose Widespread Cancer Threat," *NYT*, Sept. 26, 1992.

Page 216, par. 1: David Ehrenfeld, quoted in Stephen Viederman, "Statement to the Committee on Environmental Research, National Research Council," Jan. 15, 1992.

Page 216, par 2: Jeffery Morris and Diana Canzoneri, quote from their "Comparative Lifecycle Energy Analysis: Theory and Practice," *Resource Recycling*, November 1992, p. 26.

Chapter 10: The Business of Business

Page 218, par. 4: For a comprehensive overview of the German political-industrial system, see Kirsten S. Wever and Christopher S. Allen, "Is Germany a Model for Managers?" *Harvard Business Review*, September–October 1992, pp. 36–43.

Page 219, par. 1: On European waste reduction, see Frances Cairncross, "How Europe's Companies Reposition to Recycle," *Harvard Business Review*, March–April 1992.

Pages 218–223: For a good overview comparison of U.S. and European public policy on the environment, see David Vogel, *National Styles of Regulation: Environmental Policy in Britain and the United States* (Ithaca, N.Y.: Cornell University Press, 1986); and Ronald Brickman, Sheila Jasanoff, and Thomas Ilger, *Controlling Chemicals: The Politics of Regulation in Europe and the United States* (Ithaca: N.Y.: Cornell University Press, 1985). Brickman, Jasanoff and Ilger explain the connection between American reliance on scientific expertise and resistance to governmental authority very clearly in Chapters 7 and 8, particularly pp. 181–184, 216, 241–247.

Pages 220–221: The account of the vinylchloride regulation process derives from Joseph L. Badaracco, Jr., *Loading the Dice: A Five-Country Study of Vinyl Chloride Regulation* (Cambridge: Harvard University Press, 1985), particularly pp. 129–131, 48–55.

Page 223, par. 1: On subsidies for exploitation of forests, see Jerry Powell, "Federal Disincentives to Recycling," *Resource Recycling*, June 1992, pp. 44–45; and Donald G. McNeil, Jr., "How Most of the Public Forests Are Sold to Loggers at a Loss," *NYT*, Nov. 3, 1991.

Page 223, par. 2: Regarding energy subsidies, see the working paper of Douglas N. Koplow, *Federal Energy Subsidies: Energy, Environmental and Fiscal Impacts* (Washington, D.C.: The Alliance to Save Energy, April 1993). See also Group I of the Global Change Institute Conference on Industrial Ecology and Global Change, Snowmass, Colorado, July 20–31, 1992, p. 27.

Page 224, par. 2: The farmer is quoted in Steven Viederman, "Statement to the Committee on Environmental Research, National Research Council," January 15, 1992.

Page 225, par. 2: The discussion of the need for government support for development of environmental technologies draws heavily on George R. Heaton Jr., Robert Repetto, and Rodney Sobin, *Backs to the Future: U.S. Government Policy Toward Environmentally Critical Technology* (Washington, D.C.: World Resources Institute, 1992), especially pp. 7–8, 14–16. See also Joseph J. Rommand and Amory B. Lovins, "Fueling a Competitive Economy," *Foreign Affairs,* Summer 1993, p. 59.

Page 226, par. 1: Regarding the Ford plant, see Michael Weisskopf, "Rust Belt Emissions" *WP,* June 2, 1992.

Page 226, par. 2: On corporate culture, see John S. Hoffman, "Pollution Prevention as a Market-Enhancing Strategy: A Storehouse of Economical and Environmental Opportunities," *PNAAS* 89 (February 1992), pp. 832–834.

Page 226, par. 4: Regarding U.S. technological divestment, see Curtis Moore, "Bush's Nonsense on Jobs and the Environment," *NYT,* Sept. 25, 1992.

Pages 227–228: On the National Energy Security Act of 1991, see *Abuse of Power: Energy Industry Money and the Johnston Wallop Energy Package* (Washington D.C.: U.S. Public Interest Research Group, 1991).

Page 228–229: On U.S. investments in research, see William J. Broad, "Japan Seen Passing U.S. in Research by Industry," *NYT,* Feb. 25, 1992 and William J. Broad, "U.S. Panel Asks More for Science," *NYT,* Aug. 13, 1992; National Science Foundation, *Science and Engineering Indicators—1993* (Washington, D.C.: 1994).

Pages 229–230: On Germany's investment in the environment, I am indebted particularly to Curtis Moore, "Down Germany's Road to a Cleaner Tomorrow," *International Wildlife,* September–October 1992, pp. 24–28.

Pages 230–231: The account of Japan's environmental investment draws particularly on Douglas C. McGill, "Japan's Choice: Scour Technology's Stain with Technology," *The New York Times Magazine,* Oct. 4, 1992, p. 34; and Neil Gross, "The Green Giant? It May Be Japan," *Business Week,* Feb. 24, 1992, pp. 74–75.

Pages 231–232: On the garbage industry, see Robert C. Adler, "Financial Health and Outlook for the Waste Industry," *Waste Age,* December 1991, pp. 71–76; Timothy Egan, "Mob Looks at Recycling and Sees Green," *NYT,* Nov. 28, 1990; and Julia Flynn, "The Ugly Mess at Waste Management," *Business Week,* April 13, 1992, pp. 76–77.

Page 232, par. 2: On toxic waste cleanup expenditures, see Keith Schneider, "In Arkansas Toxic Waste Cleanup, Highlights of New Environmental Debate," *NYT,* Nov. 2, 1992; "Paying for the Past," *The Economist,* Feb. 29, 1992, p. 80; and Jessica Tuchman Mathews, "Superfund Boondoggle," *WP,* Sept. 6, 1991.

Chapter 11: Government Redux

Page 235, par. 2: Regarding the failure of U.S. executive departments to comply with environmental laws, see Keith Schneider, "U.S. Mine Inspectors Charge Interference by Agency Director," *NYT,* Nov. 22, 1992; and Matthew L. Wald, "New Disclosures over Bomb Plant," *NYT,* Nov. 23, 1992.

Page 236, par. 1: On fiscal gaps, see Michael deCourcy Hinds, "U.S. Adds Programs with Little Review of Local Burdens," *NYT,* March 23, 1992.

Page 237: par. 1: On industry slowdown on developing CFC substitutes, see Francesca Lyman, "As the Ozone Thins, the Plot Thickens," *The Amicus Journal,* Summer 1991, p. 23.

Page 237, par. 3: On conflicting priorities, see EPA (Office of Solid Waste and Emergency Response), *The Nation's Hazardous Waste Management Program at a Crossroads,* July 1990, pp. 7–9.

Page 238, par. 1: On the shortfall in EPA resources, see GAO, *Hazardous Waste: Status and Resources of EPA's Corrective Program,* April 1990; GAO, *Water Pollution: Serious Problems Confront Emerging Municipal Sludge Management Program,* March 1990; and GAO, *Toxic Substances: Federal Programs Do Not Fully Address Some Lead Exposure Issues,* May 1992, pp. 3–4.

Page 238, par. 2: On EPA rule-making delays, see GAO, *Progress in Implementing the Federal Program to Buy Products Containing Recovered Materials,* April 3, 1992; and Robert Pear, "U.S. Sets Rules to Cut Landfill Pollution," *NYT,* Sept. 12, 1991.

Page 239, par. 1: Regarding the EPA sludge program, see GAO, *Water Pollution,* pp. 16–20. On the fall in state purchasing power, see Ibid., pp. 26–27.

Page 239, par. 3: On the erosion of government statistical capability, see Jonathan Fuerbringer, "Accuracy in Short Supply in Flood of U.S. Statistics," *NYT,* Oct. 30, 1989. On EPA scientific shortfalls, see Warren E. Leary, "E.P.A. Research Lags, Report Finds," *NYT,* March 20, 1992. On EPA failure to identify risks, see GAO, *Toxic Substances: EPA's Chemical Testing Program Has Not Resolved Safety Concerns,* statement of Richard

L. Hembra, March 18, 1992, pp. 1–3. On the lack of waste site inventories, and data bases on cleanup remedies, see Keith Schneider, "U.S. Said to Lack Data on Threat Posed by Hazardous Waste Sites," *NYT*, Oct. 22, 1991; and GAO, *Superfund: Problems with the Completeness and Consistency of Site Cleanup Plans,* May 1992, pp. 44–46.

Page 240, par. 4: On EPA personnel problems, see, for example, GAO, *Superfund: EPA Could Do More to Minimize Cleanup Delays at the Clark Fork Sites,* November 1991, pp. 3, 58–59; on staff cuts under President Reagan, see *War on Waste* (see page 254), p. 67. On procurement, see GAO, *Progress in Implementing Procurement,* Washington, D.C., April 1992.

Pages 241–243: On privatization, see "Congressional Study Challenges Federal Use of Private Contractors," *NYT,* Sept. 16, 1991. On Plum Island contractors, see Keith Schneider, "Center for Animal Disease Trims Safeguards," *NYT,* March 20, 1992.

Page 242: par. 1: On the EPA data retrieval contractor, see Keith Schneider, "E.P.A. Is Called Lax with Contractor," *NYT,* Feb. 29, 1992, which cites the EPA inspector general's report of Feb. 28, 1992.

Page 242, footnote: On the EPA inspector general's estimate of questionable costs, see EPA (Office of Inspector General), *Annual Superfund Report to the Congress for Fiscal 1987,* September 1988. GAO quote on EPA failure to challenge contractor costs is from GAO, *Superfund Contracts—EPA Needs to Control Contractor Costs,* July 1988.

Page 242, par. 2: The OTA assessment is from OTA, *Assessing Contractor Use in Superfund,* January 1989, p. 40.

Page 243, par. 2: EPA officials are quoted in Keith Schneider, "Company Accused of Bilking U.S. on Waste Sites," *NYT,* March 20, 1992.

Page 244, par. 1: On California, see Dick Russell, "L.A.'s Positive Charge," *The Amicus Journal,* Spring 1991, pp. 18–23.

Page 244, par. 3: On Minnesota's packaging campaign, see Jeff Ledermann and Louise Yount, "Quantifying Packaging Waste at Grocery Stores," *Resource Recycling,* December 1992, pp. 43–46.

Page 245, par. 3: Eugene Carlson, "Small Firms Run Afoul of States' Environmental Rules," *WSJ,* Dec. 24, 1991.

Page 246, par. 1: On business's need for stability, see Keith Schneider, "Developers Leery of Wetlands Plan," *NYT,* Sept. 2, 1991.

Pages 247–250: On problems with command-and-control regulation, see Maureen Cropper and Wallace E. Oates, "Environmental Economics: A Survey," *Journal of Economic Literature,* June 1992, pp. 700–719; Sanford Gaines and Richard Westin, eds., *Taxation for Environmental Protection: A Multinational Legal Study* (New York: Quorum Books, 1991), pp. 3–5; Roger Dower and Mary Beth Zimmerman, *The Right Climate for Carbon Taxes* (Washington, D.C.: World Resources Institute, 1992), p.

69; and GAO, *Environmental Protection: Implications for Using Pollution Taxes to Supplement Regulation*, February 1993, pp. 17–19.

Pages 247–249: On regulatory modalities, see Richard N. L. Andrews, "Risk Assessment: Regulation and Beyond," in *Environmental Policy in the 1990s*, Norman J. Vig and Michael E. Kraft, eds. (Washington, D.C.: Congressional Quarterly Press, 1990), pp. 167–169; Keith Schneider, "Clinton Relies on Voluntary Guidelines to Protect Environment in Arkansas," *NYT*, April 4, 1992; Braden R. Allenby, "Achieving Sustainable Development Through Industrial Ecology," *International Environmental Affairs*, Winter 1992, p. 67; Keith Schneider, "EPA Reissues Regulations Governing Disposal of Waste," *NYT*, Feb. 20, 1992; and R. Lee Byers, "Regulatory Barriers to Pollution Prevention," *Pollution Prevention Review*, Winter 1991–92, pp. 24–25; GAO, *Environmental Enforcement: EPA Cannot Ensure the Accuracy of Self-Reported Compliance Monitoring Data*, March 1993, pp. 8–21.

Page 249, par. 2: Steven Lee Myers, "Emissions Standards Ruling Could Upset Pollution Plan," *NYT*, Jan. 20, 1993.

Page 249, par. 3: Carlson, "Small Firms."

Page 250, par. 1: On negotiated standard setting, see Kathrine Lassila, "See You Later Litigator," *The Amicus Journal*, Summer 1992, pp. 5–6.

Page 251, par. 2, 3: On Superfund costs, see Jan Paul Acton and Lloyd S. Dixon, *Superfund and Transaction Costs: The Experiences of Insurers of Very Large Industrial Firms* (Santa Monica: Rand Corporation, Institute for Civil Justice, 1992), pp. xii–xiii; also Kathleen Flood, "Why Doesn't Superfund Work?" *Waste Age*, April 1991, p. 76; and Jessica Tuchman Mathews, "Superfund Boondoggle," *WP*, Sept. 6, 1991.

Page 253, par. 3: The discussion of technology policy draws heavily on George R. Heaton, Jr., Robert Repetto, and Rodney Sobin, *Backs to the Future: U.S. Government Policy Toward Environmentally Critical Technology* (Washington, D.C.: World Resources Institute, 1992); see also Bruce Piasecki, "Industrial Ecology: An Emerging Management Science," *PNAAS* 89 (February 1992), p. 874.

Chapter 12: Policy, the Environment, and the Marketplace

Page 256, footnote: World Commission on Environment and Development, *Our Common Future* (Geneva: UNEP, 1987), p. 43.

Page 257, par. 2: On state support for recycling, see Northeast-Midwest Institute, "A Bad Burn: States Are Depending on Incineration Instead of Recycling," *Northeast-Midwest Economic Review*, Sept. 5, 1989, pp. 9–11; and Kathleen Meade, "State Recycling Budgets Show Signs of the Times," *Waste Age*, April 1992, pp. 113–116.

Page 257, par. 3: Regarding source reduction, see "Testimony by Bette K. Fishbein on the New York State Solid Waste Management Act of 1988 Before the Assembly Committee on Environmental Conservation and the Assembly Legislative Commission on Solid Waste Management," Albany, New York, September 5, 1991; and Bill Shireman, "In My Opinion . . . The Next Recycling Agenda: Growing an Economy That Reduces Waste at the Source," *Resource Recycling,* May 1992, p. 85.

Page 258, par. 1: Regarding industrial handling of toxics, see EPA, "TRI Shows Further Reductions," *Pollution Prevention News,* June 1992, pp. 1, 8.

Page 257–258: Regarding releases of toxic chemicals, see GAO, *Toxic Substances: Advantages of and Barriers to Reducing the Use of Toxic Chemicals,* June 1992.

Page 258, footnote: Regarding disincentives for recycling, see *War on Waste* (see page 294), p. 143; Jerry Powell, "Federal Disincentives to Recycling," *Resource Recycling,* June 1992, p. 44; John E. Ruston, "A Level Playing Field for Recycling Finance," *Resource Recycling,* December 1989; and Richard A. Denison and John Ruston, *Recycling and Incineration* (Washington, D.C.: Island Press, 1990).

Pages 259–263: Regarding the environmental critique of economic theory, see Herman E. Daly, *Steady-State Economics* (Washington, D.C.: Island Press, 1991); Robert Costanza, ed., *Ecological Economics* (New York: Columbia University Press, 1991). See in Costanza especially Paul Christensen, "Driving Forces, Increasing Returns and Ecological Sustainability," pp. 76–77 (on the neoclassicals and ecology); and Herman Daly, "Elements of Environmental Macroeconomics,"especially p. 33. On the environment in Samuelson and Nordhaus's textbook, see Paul A. Samuelson and William D. Nordhaus, *Economics* (New York: McGraw-Hill, 1989), pp. 118–119, 771–775, 880–883. On the reversal of scales between economy and environment, see Daly, "Elements," p. 38.

Pages 260–261: On forecasting, see Robert Repetto and Roger Dower, *Wasting Assets: Natural Resources in National Income Accounts* (Washington, D.C.: World Resources Institute, 1989), p. 1.

Pages 263–266: On environmental accounting, among many helpful sources see Jan Tinbergen and Roefie Hueting, "GNP and Market Prices," in Robert Goodland et al., eds., *Environmentally Sustainable Economic Development: Building on Brundtland* (Paris: UNESCO, 1991); Carsten Stahmer, *Cost- and Welfare-Oriented Measurement in Environmental Accounting* (Vienna: International Institute for Applied Systems Analysis [IIASA], 1991); "Wealth of Nature," *The Economist,* Jan. 18, 1992, p. 67; and Don Hinrichsen, "Economists' Shining Lie," *The Amicus Journal,* Spring 1991.

Page 265, par. 3: El Serafy quote is from his "Natural Resource Overview," in James T. Winpenny, ed., *Development Research: The Environmental Challenge* (London: Overseas Development Institute, 1991).

Pages 265–266: On European resource accounting, see Repetto et al., *Wasting Assets,* pp. 14–15.

Page 266: On revising the SNA, see Peter Bartelmus, Carsten Stahmer, and Jan van Tongeren, *Integrated Environmental and Economic Accounting: Framework for an SNA Satellite System,* Review of Income and Wealth Series 37, No. 2 (Vienna: IIASA, June 1991), particularly pp. 111, 146.

Page 267: On the extensive literature regarding estimating environmental costs and benefits, see Maureen Cropper and Wallace E. Oates, "Environmental Economics: A Survey," *Journal of Economic Literature,* June 1992, pp. 700–719; and Richard H. Westin, "Understanding Environmental Taxes," *The Tax Lawyer,* vol. 46, no. 2, pp. 337–340.

Pages 267–270: On using the market rather than regulation, see principally Robert Repetto et al., *Green Fees: How a Tax Shift Can Work for the Environment and the Economy* (Washington, D.C.: World Resources Institute, 1992); Ernst U. von Weizsäcker and Jochen Jesinghaus, *Ecological Tax Reform* (London: Zed Books, 1992); Joel Makower, "Green from the Top Down," *NYT,* Jan. 30, 1993; Christopher Flavin and John E. Young, "Will Clinton Give Industry a Green Edge?" *WW,* January–February 1993, p. 30; and Lester R. Brown, Christopher Flavin, and Sandra Postel, *Saving the Planet* (New York: W. W. Norton, 1992), pp. 144–145.

Page 268, par. 2: Regarding deposit-refund systems, see Wendy B. Pratt and Seymour I. Schwartz, "Toxic Wastes from Small Businesses Can Be Hazardous to Your Health," *Resource Recycling,* May 1992, p. 91.

Page 268, par. 3: Zosel quote is from his "Charitable Emissions Contributions?" *Pollution Prevention News,* January 1992, p. 3.

Pages 268–269: On emissions trading, see Cropper and Oates, "Environmental Economics: A Survey," pp. 687–692; Sanford Gaines and Richard Westin, eds., *Taxation for Environmental Protection: A Multinational Legal Study* (New York: Quorum Books, 1991), pp. 5–6; Roger Dower and Mary Beth Zimmerman, *The Right Climate for Carbon Taxes* (Washington, D.C.: World Resources Institute, 1992), p. 7; and Wallace Oates, "Taxation and the Environment: A Case Study of the United States," unpublished paper, University of Maryland, 1991, pp. 34, 45–6. See also GAO, *Water Polution: Pollutant Trading Could Reduce Compliance Costs If Uncertainties Are Resolved,* June 1992, pp. 2–4; and James Dao, "Some Regions Fear the Price as Pollution Rights Are Sold," *NYT,* Feb. 6, 1993.

Pages 271–277: In this discussion of environmental taxes I am deeply indebted to the analysis of Ernst U. von Weizsäcker, in *Earth Politics* (London: Zed Press, 1994); and von Weizsäcker and Jesinghaus, *Ecological Tax Reform.* See also Westin, "Understanding Environmental

Taxes," pp. 330–332, 339, 355–358; Dower and Zimmerman, *The Right Climate;* David Pearce, "The Role of Carbon Taxes in Adjusting to Global Warming," *The Economic Journal,* July 1991, pp. 938–948; and Thomas A. Barthold, "Issues in the Design of Environmental Excise Taxes," *Journal of Environmental Perspectives,* Winter 1994; Repetto et al., *Green Fees,* Flavin and Young, "Green Edge,"

Page 271, par. 3: On losers, see von Weiszäcker and Jesinghaus, *Ecological Tax Reform,* p. 64. Popoff quote is from his "Life After Rio: Merging Economics and Environmentalism," a paper presented at the Chemical Week Conference, An Environment of Change, Houston, Oct. 15, 1992.

Page 272, par. 3: On revenue neutrality, see von Weizsäcker and Jesinghaus, *Ecological Tax Reform,* p. 18; Pearce, "Role of Carbon Taxes," pp. 940–941; and Dower and Zimmerman, *The Right Climate,* pp. 8–10, 13–15.

Page 273, par. 1: On gradual application of taxes, see von Weizsäcker and Jesinghaus, *Ecological Tax Reform,* p. 24; also von Weizsäcker, *Earth Politics,* p. 133; Pearce, "The Role of Carbon Taxes," pp. 941–942; Dower and Zimmerman, *The Right Climate,* p. 20; and Westin, "Understanding Environmental Taxes," p. 339.

Pages 273–274: On phasing out of petroleum, see von Weizsäcker, *Earth Politics,* pp. 31–32, 51–52.

Page 274, par. 3: On EPA estimates on a carbon tax, see William K. Stevens, "New Studies Predict Profits in Heading Off Global Warming," *NYT,* March 17, 1992. On new demand, see von Weizsäcker and Jesinghaus, *Ecological Tax Reform,* p. 80.

Page 274, par. 4: On the relationship between high energy prices and economic performance, see von Weizsäcker and Jesinghaus, *Ecological Tax Reform,* pp. 80, 71–72. On Japan, see Council on Environmental Quality report on Japan, 1990, pp. 229–230. On stable incentives for environmental investments, see Pearce, "The Role of Carbon Taxes," pp. 941–942.

Page 275, par. 2: On unintended consequences, see Robert A. Frosch, "Industrial Ecology: A Philosophical Introduction," in *PNAAS* 89 (February 1992), p. 803.

Page 275, par. 3: On the regressivity issue, see Dower and Zimmerman, *The Right Climate,* pp. 21–28; Westin, "Understanding Environmental Taxes," p. 334; and Pearce, "The Role of Carbon Taxes," p. 943.

Page 276, par. 3: On pollution prevention and command and control regulation, see Frederick R. Anderson, "Regulatory Pollution Prevention" (paper prepared for Woods Hole Conference on Industrial Ecology, July 1992).

Chapter 13: Progress and Plenty in the Twenty-first Century

Page 280, par. 3, 4: Donella F. Meadows, *The Limits to Growth* (New York: Universe Books; 1972). See also Paul Kennedy, *Preparing for the Twenty-first Century* (New York: Random House, 1993), pp. 21–46.

Page 288, par. 3: On IBM, see Barbara S. Hill, "Industry's Integration of Environmental Product Design," Proceedings of the Institutes for Electrical and Electronics Engineering (IEEE), May 1993; also, Bette K. Fishbein, *Germany's Packaging Law: A Nation Confronts Its Solid Waste Crisis* (New York: INFORM, 1994), p. 27. On Xerox, see Jack Azar, "Asset Recycling at Xerox," *EPA Journal,* July–September 1993, pp. 14–16; also Fishbein, *Germany's Packaging Law,* pp. 27–28.

Page 289, par. 4: On the service economy, see Orio Giarini and Walter Stahel, *The Limits to Certainty: Facing Risks in the New Service Economy* (Dordrecht, Germany: Kluwer Academic Publishers, 1989). I am indebted to this imaginative analysis, particularly regarding risk (see pp. 37–39, 52); delivery systems (see pp. 2–4); software and maintenance (see pp. 32–33); the need for performance standards (p. 47); in-situ monitoring (pp. 54–55); repair prices (p. 54); and small workshops (p. 51).

Index